The Structure and Dynamics of Human Ecosystems

The Structure *and* Dynamics *of* Human Ecosystems
Toward a Model for Understanding and Action

WILLIAM R. BURCH, JR.
GARY E. MACHLIS
JO ELLEN FORCE

Yale UNIVERSITY PRESS/NEW HAVEN & LONDON

Published with assistance from the foundation
established in memory of Philip Hamilton McMillan
of the Class of 1894, Yale College.

Copyright © 2017 by William R. Burch, Jr.,
Gary E. Machlis, and Jo Ellen Force.
All rights reserved.

This book may not be reproduced, in whole or in part,
including illustrations, in any form (beyond that copying
permitted by Sections 107 and 108 of the U.S. Copyright Law
and except by reviewers for the public press), without written
permission from the publishers.

Yale University Press books may be purchased in quantity for
educational, business, or promotional use. For information,
please e-mail sales.press@yale.edu (U.S. office) or
sales@yaleup.co.uk (U.K. office).

Set in Minion type by Newgen North America.
Printed in the United States of America.
Library of Congress Control Number: 2016962156
ISBN 978-0-300-13703-3 (cloth : alk. paper)

A catalogue record for this book is available from
the British Library.

This paper meets the requirements of ANSI/NISO Z39.48–1992
(Permanence of Paper).

10 9 8 7 6 5 4 3 2 1

Contents

Preface vii
Acknowledgments ix

ONE Introduction 1

TWO An Overview of the Model 12

THREE Lessons and Legacies 23

FOUR The Ecosystem Concept in Biology 51

FIVE The Roots of Human Ecology 64

SIX Key Components and Variables for Analyzing Human Ecosystems 74

SEVEN Goals, Strategies, and Tactics for Inquiry and Action 102

EIGHT Using the Model for Science during Crisis 133

NINE Revitalizing Human Communities and Reclaiming
Biological Communities: The Baltimore Story   148

TEN Toward a More Perfect Civic Order:
Lessons Learned from Research   168

ELEVEN Extending the Capability of the Model   210

TWELVE Leaning Forward:
Future Challenges to Human Ecosystems   240

THIRTEEN Conclusion   267

Notes   271
Index   293

Preface

We often hear that humanity has made a Faustian bargain in consuming beyond the Earth's carrying capacity and that our arrogance as a species means that we are ultimately doomed to a hellish future. There is no question that we live in a new Anthropocene world that provides grave challenges. We have extraordinary appetites of consumption and powers of destruction. Yet, what often gets forgotten is that we have equally sophisticated skills of rehabilitation, adaptation, and transformation. Ecosystem conditions from local to global require new tools for decision-making, new decisions, and (perhaps) new decision-makers. The application of scientific knowledge to environmental problems is fraught with perils and possibilities.

Our choice is between two Faustian tales. There is the late sixteenth-century dramatist Christopher Marlowe's Faustian tragedy, in which Faust makes a bargain with the Devil and his soul is taken by the Devil. His last words renounce the accumulation of knowledge: "Ugly hell gape not! Come not Lucifer! I'll burn my books!" The other version was published as a play, *Faust,* in two parts (the first in 1808, the second in 1882) by the scientist, writer, and artist Johann Wolfgang von Goethe. It is not a contract with the Devil but a wager between God and the Devil. At the end the Devil is about to take Faust's soul to hell, but the Devil is distracted by God's cherubic angels. Faust is raised to heaven, where his redemption is due to Faust's continual striving: "He who strives on and lives to strive can earn redemption still."[1]

Marlowe's story reveals a selfish humanity that turns its back on knowledge; Goethe's *Faust* presents hope for redemption for our failures through continual learning and striving to apply what we have learned. We can learn to avoid actions that destroy. We can learn to reclaim and to sustain ecosystems that have been abused, and cities can be made pleasant and functional. This meaning is the key for application of the Faustian story to contemporary environmental issues. It is Goethe's Faustian wager that we embrace, and that is why this book was written.

Acknowledgments

A years-long project such as this gathers significant intellectual debts along the way. We wish to acknowledge and sincerely thank the hundreds of students around the world, many now mid-career professionals, who have learned about and used the Human Ecosystem Model and its concepts and helped us develop and improve the model over the decades. Graduate students at Yale University and the University of Idaho have been tireless data collectors, astute critics, and superb information and literature seekers in libraries, courthouses, newspaper and government archives. We thank the numerous resource managers, community leaders, scientists, and citizens who have participated in workshops, discussion groups, project teams, and seminars that have helped hone and polish the model.

In addition, we acknowledge the intellectual and spiritual debt we owe several individuals. Professor F. Herbert Bormann helped us see how ecosystem ecology could be used as a foundation for social sciences knowledge. Dr. Ralph Jones encouraged us to attempt the community-based approach being explored in Nepal and Bhutan and to bring it to the struggling American city of Baltimore. Dr. Ram Guha's work with the Chipko people demonstrated that respecting the dignity of local knowledge is critical for understanding human ecosystems. Dr. Morgan Grove, our long-term colleague, has provided wise counsel and sharp insights as our work and the human ecosystem model evolved over time.

We are deeply grateful to Alexis Ward at Clemson University. Alexis prepared innumerable drafts and responded to our (seemingly) endless revisions with good cheer and high skill. We also thank Martin Schneider for his skilled and thoughtful editing.

We began this work toward the end of the twentieth century and share it with readers now at the early stages of the twenty-first. Sometimes books burst forth, written quickly. Others, like this one, emerge slowly over time. We thank Yale University Press for its support and patience.

The errors are ours, readers; judgment is yours.

O · N · E

Introduction

1

Major environmental events are challenging people and communities in general and the environmental professions in particular. There is little question that, around the world, population growth coupled with sharply rising human wants and needs will intensify ecological demands. Complex, difficult, and often unintended consequences are likely. Confronted with the dilemma of managing human uses of biophysical resources— from transportation planning in urban Tokyo to restoration efforts on Connecticut's Housatonic River to wildlife management in Idaho to the conservation of biodiversity in the Galápagos archipelago—how can the manager, scientist, student, policy-maker, and citizen respond?

Population growth creates burgeoning environmental challenges —this has, of course, become a commonplace pronouncement since Paul Ehrlich's 1968 warning *The Population Bomb*. Yet it is far from banal. Around 2025, the globe is predicted to carry eight billion humans and by 2050 the total population could reach 9.6 billion persons. Almost all of this population growth is in developing countries, especially in Africa.[1]

In contrast, consumption patterns reflect the industrial exuberance, strong appetites, and market share of the developed world. A quarter century ago, the developed world's 25 percent of global population "lay[s] claim to 85 percent of all forest products consumed, 72 percent of steel production, and 75 percent of energy use. Developed countries also generated about 75 percent of the global burden of pollutants and wastes."[2] The distinction between "developed" and "developing" countries is currently less clear. The metric that is often used to indicate the relationship between consumption and production of resources and pollutants is greenhouse gas emissions. China now emits the largest volume of greenhouse gases annually, followed by the United States, the European Union, and India. On a per capita basis, Canada leads with 25 tons CO_2/capita. China is only slightly above the world average of approximately 7 tons per capita, and India is only one-third the global average.[3]

Yet it is not only population growth, the contemporary means of production, and greenhouse gases that create our robust environmental challenges. The "structures of distribution"—including economic, social, and political arrangements worldwide—are also critical. Economist Lester Thurow is stereotyping but insightful: "If the world's population had the productivity of the Swiss, the consumption habits of the Chinese, the egalitarian instincts of the Swedes, and the social discipline of the Japanese, then the planet could support many times its current population without privation for anyone. On the other hand, if the world's population had the productivity of Chad, the consumption habits of the United States, the inegalitarian instincts of India, and the social discipline of Argentina, then the planet could not support anywhere near its current numbers."[4]

The magnitude and consequences of this imbalance are described (with many data and heated prose) in Paul Ehrlich and Anne Ehrlich's *One with Nineveh: Politics, Consumption, and the Human Future*. According to the authors, each baby born in the United States will, on average, "cause" 15 to 150 times more environmental damage than a baby born in a "very poor country." Americans, desirous of "tract mansions," enmeshed in a "berserk car culture," and urged into "overconsumption," have a wide and destructive footprint. Developing nations with their new consumers and new consumption patterns are also targeted: "China, which has more new consumers (over 300 million) than the

United States has consumers (slightly less than 300 million), has virtually doubled its per capita meat consumption since 1990, making it the world's largest carnivorous nation (the power of some 1.3 billion caputs!)."[5]

Consequently, environmental scientists, managers, and other professionals will need to intensify their search for cross-cultural models of resource systems that include the forces driving human desires. Sociocultural variables as both cause and consequence of system change will need to be joined to the traditional biophysical concerns of the ecologist, forester, conservationist, range manager, park superintendent, and rural agriculturalist. Urban ecosystem managers—from technocrats managing the sewer system of Kuala Lumpur, to city planners in Alexandria (Egypt or Virginia)—will also require interdisciplinary models that can help them to manage potable water, energy, urban open space, food supplies, city transportation, and so forth.

The challenges are at once localized and global: geochemical flux due to pollution, the accumulating inventory of toxic chemicals, a collapsing world fishery, biotic mixing (that is, the introduction of exotic species), accelerating loss of coral reefs, desertification, and more. Gus Speth, in his 2004 book *Red Sky at Morning,* lists nine "interacting drivers of biotic impoverishment: land use conversion, land degradation, freshwater shortages, watercourse modifications, invasive species, overharvesting, climate change, ozone depletion, and pollution."[6]

Worldwide, disparate indicators of ecosystem stress abound. Caribbean coral reefs exhibit dramatic loss measured by coral cover (reduced up to 80 percent), reduced resilience and functional species groups, and reef erosion at rates of up to 10 $kgm^{-2} yr^{-1}$. Within the protected areas of Indonesian Borneo, lowland forests have declined by more than 56 percent (more than 29,000 square kilometers) during the 1985–2001 period. Based on several decades of comprehensive surveys of plants, birds, and butterflies in Great Britain, an estimated "28 percent of native plant species have decreased over the past 40 years . . . 54 percent of native bird species have decreased over 20 years, and . . . a majority of butterfly species (71 percent over ~20 years) has declined." The authors dryly note: "Population extinctions were recorded in all the main ecosystems of Britain, and were distributed with remarkable evenness across the nation, rather than concentrated in a few degraded regions."[7]

There is a sense of desperation in many of these pronouncements of doom. The Ehrlichs' grim metaphor of the abandoned city of Nineveh avoids the actual and more hopeful biblical result—the people of Nineveh repented and were saved from destruction. Marlowe's Faustian surrender is echoed by Robert M. May when he warns of species extinction: "It seems a pity to be burning the books before we can read them, and before we can create wealth from the recipes on their pages."[8]

Yet, whichever way one reads the data and trends, all but the most blinded of ideologues will accept that ecologically, these are challenging times.

2

How can the manager, scientist, student, policy-maker, and citizen respond? We note the suite of too-common responses: onset of disciplinary myopia, reliance on trained incapacities, cynical delusion, the search for single-variable causes or xenophobic fault, and linear projections of the future based on often-flawed assumptions about the past and present. These responses are unlikely to be successful, given the complexities of nature and of our species in particular. *Homo sapiens* are not so fortunate as to have their ecological challenges be separate, distinct, and amenable to simple solutions. Real-world complexity demands interdisciplinarity.

This book is an effort to provide and describe one possible tool—the Human Ecosystem Model (HEM). We strive to describe formally the structure of human ecosystems through this interdisciplinary model and to show how to apply it as a form of usable knowledge. The model is built using selected and defined variables, and we demonstrate how it can be tested, advance scientific inquiry, serve resource management, inform environmental decision-making, and reflect further improvements.

In the United States, the Deepwater Horizon disaster in the Gulf of Mexico involved a sea-floor oil gusher that flowed for eighty-seven days from April 20 until July 15, 2010. In late October 2012, the United States was faced with another major disaster when Hurricane Sandy moved up the entire eastern seaboard from Florida to Maine and westward as far as Wisconsin, affecting twenty-four states, with particularly severe damage in New Jersey and New York.

As each of these events unfolded, the Department of Interior established a Strategic Sciences Group (SSG) to assess how the event might impact the ecology, economy, and people in the regions affected. Each SSG included scientists from diverse disciplines and from federal, academic, and nongovernmental organizations. The work was intended "to provide rapid scientific assessment of potential consequences of the [disaster] that could provide useable knowledge to decision-makers."[9] The Human Ecosystem Model presented in this book guided efforts of the SSGs by providing a conceptual map that helped to generate a series of cross-disciplinary questions and scenarios for anticipating and monitoring future likely events. Clearly, there are cascading effects, but the empirical connections between biophysical and sociocultural variables remain unclear. We believe that the HEM can provide guidance to policy and decision-makers, researchers, and students in future planning, management, and environmental and natural resource situations and challenges.

We admit that our interest in both the specific event (like Hurricane Sandy) and the universal, that is, regularities across cultures, times, people, and places. Such regularities feature prominently in our model, for they (along with the intriguing pleasures of biophysical and sociocultural idiosyncrasies) allow a generalized understanding of human behavior and its consequences. We share with others a fascination with the challenge to document, as the anthropologist Sydney Mintz describes, "the marvelous variability of human custom while vouchsafing the unshakable essential oneness of the species."[10]

Consider, for example, food—universally important (besides sex, what else so dominates human waking hours?) and deeply dependent on ecosystem functioning. In *Near a Thousand Tables: A History of Food,* Felipe Fernández-Armesto cheerfully notes that "our most intimate contact with the natural environment occurs when we eat it." Cooking—a particularly human practice—has its regularized technology, economy, cultural meaning, and nutritional consequence. Fernández-Armesto describes (through basic chemistry) its importance: "The cooking revolution was the first scientific revolution: the discovery, by experiment and observations, of the biochemical changes which transform flavor and aid digestion.... Cooking makes the proteins in the [meat] muscle fibers fuse, turning collagen to jelly. If direct fire is applied, as was probably the case in the earliest cooks' techniques, the surface of the meat undergoes something like caramelization as the juices are concentrated: for proteins coagulate when heated and 'Maillard reaction' sets in between the amines on a protein chain and some of the natural sugars in fat.... Heat retextures other foods so that they can be chewed or easily dissected by hand—a primary spurt in the civilizing of eating habits, long before the introduction of chopsticks or knives and forks."[11]

Food is a universal; certain forms of food preparation are nearly so: The American suburban barbeque is a ritualized descendant of the banquet feasts of Nestor in the *Odyssey*. At the same time, strategies for eating and rules for what to eat are wildly varied: "Americans eat oysters but not snails. The French eat snails but not locusts. The Zulus eat locusts but not fish. The Jews eat fish but not pork. The Hindus eat pork but not beef. The Russians eat beef but not snakes. The Chinese eat snakes but not people. The Jale of New Guinea find people delicious."[12]

The production, distribution, consumption, and elimination of food are all embedded in the real life of human ecosystems, as are the ritualized meaning, display, taboo, and use of food as symbol. Other behaviors—the ways we work, make love, fight, play, and so on are also a mixture of the universal and the culturally specific, the global and the localized, the material and the symbolic. To fully understand *Homo sapiens*—and human ecosystems—the regularity *and* variety of human activity must be contrapuntally considered. Biophysical and sociocultural traits are simultaneously critical.

3

Our intent is to outline the structure of human ecosystems on the scale of "grand theory" (without apology, and, in fact, with enthusiasm) for both reasons of strategy and interest. The search is to describe ecological organization, pattern, and process within the contemporary ecology of our species and to apply this description in practical ways.

Our intended audience includes the manager, scientist, scholar, student, and policy-maker. It generally includes those interested in environmental affairs, the integration of the natural and social sciences, and the advance of insights from systems ecology into the realms of human behavior. And we write for several specific audiences as well.

For the resource manager—the forester, water manager, urban planner, park ranger, and the like—our brief argument provides a useful tool for inventory, analysis, strategy, and action. For the scientist and scholar, our use of the human ecosystem as an organizing concept, and the specific model proposed in the following pages, is worthy of consideration, testing, and improvement.

For the student—primarily graduate students in professional training for environmentally oriented careers—the book may be a source of inspiration or at least creative irritation, as well as a chance to synthesize the sciences in ways that reveal new possibilities. For the policy-maker, our book is again a potential tool for insight, strategy, and action—and a primer on the human ecosystem as an organizing concept.

For all willing readers, we will attempt to envision future application of the human ecosystem concept to both the "critical synthesis" of the natural and social sciences, and to the environmental decisions confronting our species in the twenty-first century. The link between theory and practice is paramount. An example is the emerging socio-technical process of "ecosystem management."

4

Since 1990, ecosystem management (or more generally, the ecosystem approach) has carried substantive hopes for finding some coherent and comprehensive means of systematically fitting human demands within biophysical and sociopolitical realities. There are numerous definitions of ecosystem management and also vigorous debate as to its implications. Although most of the literature on ecosystem management is from the 1990s, Susan Leech and her colleagues provide a summary of some definitions from the literature as follows: "The basic tenet of an ecosystem-based approach is that conserving ecosystem functions and integrity is vital because viable ecosystems are the basic life support system for human communities."[13]

The organization and description of comprehensive models for ecosystem management are evolving processes. Strategic inclusion of people in such efforts is now commonly touted and occasionally creatively explored.[14] The full substantive integration of human variables into ecosystem models is as yet unrealized.

Biologists have focused on "impact" measures of humans such as harvest rates, caloric consumption, and habitat fragmentation. It is a strategy that puts our species outside the ecosystem as, at most, a permanent perturbation. Social scientists have largely focused on idiosyncratic "human dimensions" but are sometimes wistfully immune to biological reality. The traditional academic divisions have worked at intellectual balkanization, seeking advances in territory rather than a more inclusive paradigm that would be truly helpful to resource management professionals. Such a paradigm would be kin to a new kind of applied life science, one that treats the biosocial reality of human beings as a core element of its approach toward ecosystem management. We are reminded of Peter Heylyn's wise warning concerning displacing barriers, penned in his *Cosmographie*: "[History and Geography], if joined together, crown our reading with delight and profit; if parted, [they] threaten both with a certain shipwreck."[15]

We will attempt to provide a viable and practical synthesis and through it describe the structure of human ecosystems. As we hope to demonstrate, the structure of human ecosystems is fundamentally dynamic and includes parts, flows, patterns, and processes—all of which are simultaneously biophysical and sociocultural.

We have chosen deliberately to treat the structure of human ecosystems as comprehensive of organizational form and process functioning. The structure of urban centers such as Baltimore and Ho Chi Minh City is embodied in the component streets, buildings, and parks *and* in the flow of people in and among these parts of the city. The structure of rural water delivery systems in South Africa is embedded in pipeline but also water flow rates, demand for water by time of day, cultural rules for cooking, engineered artifacts of apartheid, and modern management practices. In human ecosystems, process *is* part of structure.[16]

In addition, we follow Stephen Jay Gould's analysis of intellectual "structure" as the parsimonious construction of scientific theory: "shared content, not only historical continuity, must define the *structure* of scientific theory; but this shared content should be expressed as a *minimal list* of the *few defining attributes* of the theory's *central logic*—in other words, only the absolutely essential statements, absent which the theory would either collapse into fallacy or operate so differently that the mechanism would have to be granted another name."[17]

The structure of human ecosystems to be described and applied in this book follows Gould's remonstrance: a minimal list of the defining attributes of the model's central logic. (Of course, the late Dr. Gould knew ironic potential when he saw it—his masterwork *The Structure of Evolutionary Theory* takes 1,433 pages to make its argument.)

We will propose a key agency or "driver" that leads to human ecosystem change, treating adaptation in response to human needs and wants as central. We will attempt to provide a testable set of proposals regarding the pattern, process, and agency of human ecosystems, along with methodological alternatives for testing and improving our initial effort.

The model should help answer the question posed earlier: Confronted with the dilemma of managing human uses of biophysical resources—how can the manager, scientist, student, policy-maker, and citizen respond?

5

The book is organized as follows. In the next chapter we describe the HEM in its present form. Chapter 3 describes the evolution of the model over nearly a half-century of work by the authors. The legacies that this initial work built on, and the lessons learned through collaborations and applications with students and colleagues across the world and over the decades, tell a collective story of striving, erring, and striving again in the Faustian tradition. Chapters 4 and 5 continue with an intellectual genealogy of the ecosystem concept and the roots of human ecology.

Chapter 6 returns to the HEM and provides more detailed information about the essential components and variables of the model. Chapter 7 presents some guidance for the user to develop strategies and tactics when using the model, including some examples of a how the HEM is being applied to real-world ecosystems. In chapter 8, we present the most recent use of the HEM as a guiding framework for the use of science during an environmental crisis (the Deepwater Horizon oil spill in the Gulf of Mexico) and a natural disaster (Hurricane Sandy). Use of the model was of interest to scientists, managers, and policy-makers as they faced the challenges resulting from these disasters. Chapters 9 and 10 focus on an elaborate and ongoing application of the model in Baltimore, Maryland.

We then turn to possible future directions of human ecosystems and the capacity of the HEM to serve changing needs. First, we offer some of our thoughts on how better to link human and biophysical factors, to develop better predictable capability, and to adapt to large, densely populated urban ecosystems. We then look to the challenges going forward in the ecology of *Homo sapiens*. We speculate on some often overlooked but significant trends in social institutions, with emphasis on justice, sustenance, and faith institutions. There is a consistent yet still emerging realization that interdisciplinary approaches are required, if complex environmental problems are to be constructively confronted and the human condition improved. Steadily, the blinders of exclusive disciplinarity are being lifted and replaced with tentative forms of interdisciplinary collaboration.

Yet what we are striving for is perhaps bolder than a tentative and incremental synthesis. We argue for a new applied "life science" capable of creating a full natural history of contemporary *Homo sapiens*. We believe that such a life science will emerge as one that treats the concepts, theory, methods, and findings of the natural and social sciences with mutual curiosity, enthusiasm, respect, and learning. There is a real possibility for such a new and emergent science—one that integrates natural and social systems with fidelity to scientific methods, appreciation of the joy of discovery, and duty to the need for creating usable knowledge.

In the last chapter, we conclude our brief argument, with its particular attempt and application. It is, of course, a mere and modest stepping stone on any path toward this end. The Faustian bargain of Goethe's *Faust* provides for human labors: We strive, err, and strive again. Such efforts, no matter how incomplete, are ultimately redemptive. There is much at stake, for the fate of human ecosystems is our own.

T · W · O

An Overview of the Model

1

In this chapter, we describe our Human Ecosystem Model (HEM), which began as a seedling in the late 1960s and has grown and developed into a middle-aged forest today. We know that just as real-world ecosystems evolve and adapt into a variety of dynamic human and "natural" ecosystems existing on the planet, our HEM will also evolve and adapt as it is used by us and others. Some ecosystems, like some models, are resilient and sustainable, and some are not. We have strived to develop a model that can guide research that contributes to our scientific understanding of resilient ecosystems and that can guide policy-makers as well as resource managers and citizens toward the goal of a sustainable and superior quality of life for all the planet's creatures. We define the human ecosystem as *a coherent system of biophysical and social factors capable of adaptation and sustainability over time.*

Coherence implies (a) a unity of specific functions or processes, and/or (b) the interrelatedness of a reasonably identified whole or unit (this varies from the term's use in physics, where coherence describes a constant phase relationship). Coherence is relative and varying within human ecosystems. There can be "tight coherence" between available energy flows and economic activity, as when a scarcity of petroleum reduces industrial production and raises costs. There can also be "loose coherence," such as the relationship between available water supplies and the performance of educational institutions. This relationship is loose but existent. For example, schools need water for students; in many parts of the world, the lack of nearby water sources can force persons (most often women) to allocate significant time to transporting water supplies to the school, which may reduce opportunities for the education of women.

Coherence is a measurable quality critical to the identity of human ecosystems as systems. While human ecosystems are by definition *systems,* they are systems of a particular kind. First, they are open systems rather than closed. Human ecosystems have significant external inputs (from sun-driven energy to capital transfers) and similarly significant outputs (from pollution to profits). Like coherence, the degree of openness can be measured for each human ecosystem and is an essential characteristic of the system. Second, human ecosystems are organic rather than mechanical. We do not mean the vague organicism of the early ecologists but invoke the realization that such systems are "life systems" and hence stochastic, nonlinear, and subject to chaos-induced change and fractal-like complexity. The machine metaphor for human systems is both an accessible starting point and an eventual frustration to understanding how human ecosystems actually operate. Third, human ecosystems are evolutionary, but not in a Darwinian sense. That is, they can evolve over time, but in ways and through means different from the mechanisms of evolutionary biology described by Darwinian theory. This evolutionary trajectory has significant implications for understanding the processes and consequences of ecosystem change.

Our definition of human ecosystems identifies specific *biophysical and social factors* of importance. The key *elements* are critical resources (including biophysical, socioeconomic, and cultural resources) and the social system. It would be a misreading to treat these two elements as fully discrete and separate categories or kinds. Within these two key elements, there are several *components,* which include a set of similar *variables* within each component. For example, within the component *cultural resources* there are the variables *organization, beliefs, arts, crafts,* and *myths.* Within the component *cycles,* there are four variables (or types of cycles): *physiological, individual, institutional,* and *environmental.*

Capability ("a system . . . capable of adaptation") implies both the contingency and potential of human ecosystems. We do not suggest that human ecosystems exhibit static balances, constant and always effective adaptation to external forces, or other idealized and unrealized harmonies. We do suggest that human ecosystems will exhibit regular though varied effectiveness at adaptation. Many will exhibit at least short-term sustainability—commuters commonly arriving at their destination, meals served at dinnertime, crops brought to market, gardens tended, monies reliably exchanged, seasons celebrated, and so forth. The contingency of human ecosystems is expressed when these patterns are interrupted—the small and normal accidents of commuters delayed, crops spoiled, gardens abandoned, monies lost, celebrations cancelled, and so forth. It is the steady mix of regularized patterns and periodic contingencies that creates the variability of human ecosystems and the need for adaptation.

Adaptation is used here in a non-valued sense to reference a "response to conditions." Farmers plant a different grain in response to less or more rain, labor, money, fuel, or soil. Schoolchildren may carry an extra coat (or, unfortunately, a weapon) to deal with school conditions. Adaptation is ubiquitous in human ecosystems. Hence, adaptation is itself a base condition of human ecosystems. It is also contingent: What is adaptive (or advantageous) for one person, institution, or social group may be maladaptive (or harmful) to another person, institution, or social group.

Sustainability, as used in our definition of human ecosystems, has both sociocultural and biophysical meanings. Ernest Callenbach, who penned the classic *Ecotopia* in 1975, provides an elegant description: "From a human point of view, a sustainable society is one that satisfies its needs without diminishing the prospectus of future generations. . . . From a rigorous ecological point of view, an *ecosystem* operates sustainably if its inputs and outputs (of both energy and materials) are balanced; over time it is not losing substantial amounts of nutrients. Such a situation can be described as dynamic equilibrium or a 'steady state,' although there are always fluctuations at work."[1]

In our definition of human ecosystems, the phrase "over time" presents a particular dilemma. Time frames vary dramatically by human purpose, culture, historical epoch even socioeconomic, ethnic, or national categories. How long does a "coherent system of biophysical and social factors" need to be present before it fits our definition of a human ecosystem? When does a human ecosystem so alter its assembly of parts, flows, patterns, and processes that it diminishes to a sub-minimal coherence and/or becomes something else? We will take up these questions again in later chapters, as have the biological ecologists concerned with understanding succession, exotic systems, and ecosystem restoration. For the present, the key point is that "over time" is always relative, culture-bound, and important to specify.

To summarize: The human ecosystem is defined as *a coherent system of biophysical and social factors capable of adaptation and sustainability over time.* A rural village in western Russia or California Baja Sur, Mexico, can be considered a human ecosystem if it exhibits identifiable boundaries, essential ecosystem functions, resource flows, social structures, social processes (including adaptive responses to changed conditions), and dynamic continuity over a period of time. So can the villages of East Africa and the great mega-cities of Asia, Europe, and North America.

Hence, human ecosystems can be described at several spatial scales, and these scales are hierarchically nested. A family household, local neighborhood, watershed, community, province, region, nation-state, and even the planet can be treated as a human ecosystem. To do so requires an explicit and broad description—and a reasonably comprehensive conceptual model of essential components, variables, and flows. Such a model would cumulatively describe the structure and dynamics of human ecosystems and be potentially applicable to a wide range of places and persons.

2

We now expand on the parts of our Human Ecosystem Model, which is presented in figure 2.1. A brief tour is useful. First, we distinguish among *constants, base conditions,* and *variables.* Constants do not measurably vary and are universal; thermodynamic properties are examples. Base conditions exhibit little (but some) variation over space and time and

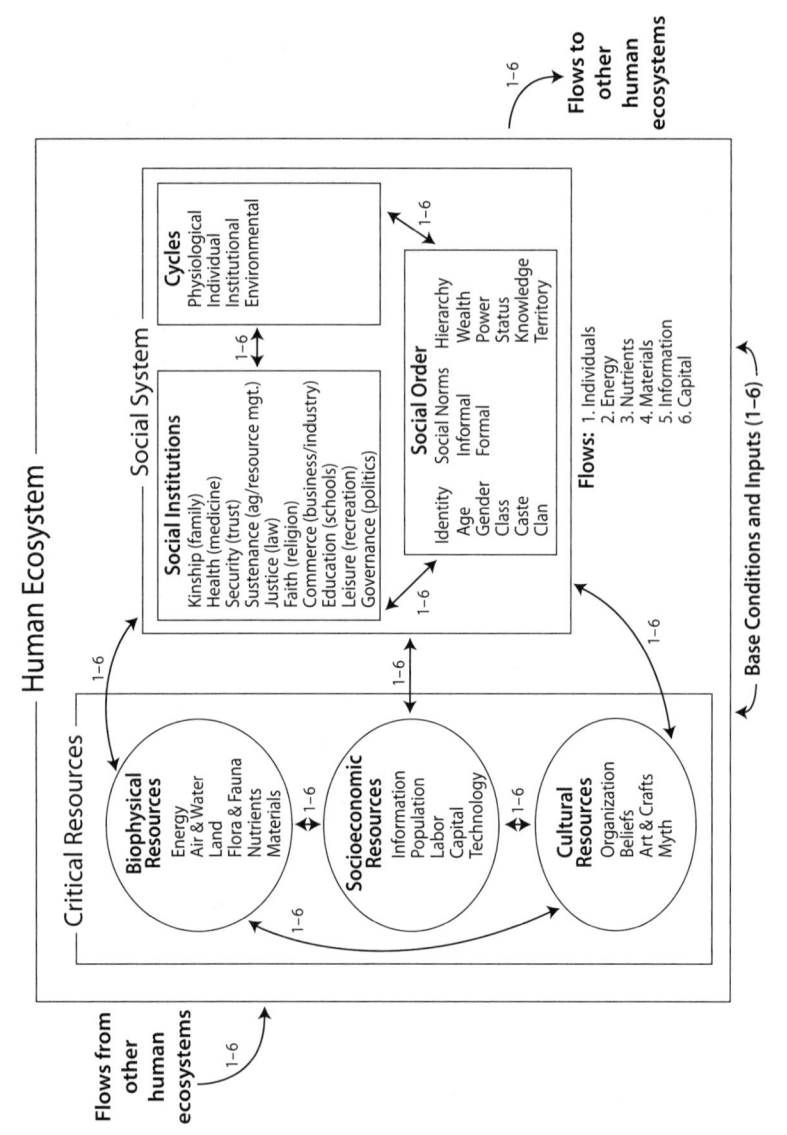

Figure 2.1. The Human Ecosystem Model

are ubiquitous in human ecosystems; the presence and flux of the nitrogen cycle is an example. Variables are those factors that potentially fluctuate widely in qualities over space and/or time; actual fluctuations may be less.

Human ecosystems rest on a foundation of abiotic and biotic factors taken as *base conditions* and inputs. These base conditions include a solar-driven energy source obeying thermodynamic laws, biogeochemical cycles of great constancy, land forms and geological variation of significant complexity, and the full genetic structure of life including the genetic and biophysical properties of *Homo sapiens.*

Vaclav Smil, in an extraordinary three paragraphs, summarizes the immensity and implications of base conditions:

> As long as the *sun* has enough hydrogen in its core to sustain orderly thermonuclear reactions, the star will flood the *Earth* with a surfeit of *solar radiation* that will continue to energize most of the physical and chemical processes on our planet. This *radiation* heats the *atmosphere* and the *ocean,* generates *winds* and *rains,* and powers the inexorable advances of surface *denudation.*
>
> Sunlight also energizes *photosynthesis,* the planet's most important biochemical conversion, creating new biomass in *bacteria, phytoplankton,* and higher *plants,* and above all in *forests* and *grasslands.* This synthesis is the foundation of *food chains* providing the nutrition needed for the *heterotrophic metabolism* of *animals* and *people,* nutrition that allows activities ranging from simple *running* to elaborate tasks of *labor and leisure.*
>
> Human societies—from small groups of *hunters and gatherers* to complex societies dependent on enormous flows of *fossil fuels and electricity*—have all been inextricably tied to the steady flux, and converted stores, of sunlight. Accumulated stores of photosynthetically converted solar radiation, tapped as *coals, crude oils,* and *natural gases,* will continue to serve for many generations as the foundation of *fossil-fueled*

civilization with its profusion of energy services, ubiquitous transportation, and surfeit of *information*.[2]

Base conditions include constants—the formidable first and second laws of thermodynamics being paramount. Because of the first law of thermodynamics, energy is never consumed; in the universe it is a constant and has an absolute limit. Since by the first law, energy cannot be consumed, how can energy actually be "used"? The answer is provided by the second law of thermodynamics, which states that in any energy-based process, some energy is always degraded in quality, so that the ability to work is lessened. As Smil notes, "The second law of thermodynamics addresses the inescapable reality that the potential for useful work steadily diminishes as we move along energy conversion chains. There is a measure associated with this loss of useful energy, and it is called *entropy*. While energy is conserved in any conversion, the conversion can only increase the entropy of the system as a whole. There is nothing we can do about this decrease of utility. A barrel of crude oil is a low-entropy store of very useful energy that can be converted to heat, electricity, motion, and light."[3]

Base conditions also include key biogeochemical cycles, which create both planetary fluxes and fine-scale biological activity. Table 2.1 provides a summary of several but not all of these important cycles: water, nitrogen, carbon, sulfur, and phosphorous being examples.[4]

The full inventory of base conditions and their attendant inputs and processes is both lengthy and instructive. Any effective model of

Table 2.1. Selected biogeochemical cycles as base conditions

Cycle	Selected Key Function(s)
Water	Basic life processes
Nitrogen	Cell proteins and gene functioning
Carbon	Cell functioning and atmospheric regulation of temperature
Sulfur	Cell functioning and regulation of global temperatures
Phosphorous	Cell membranes, gene functioning

Source: *Adapted from Callenbach*, Ecology, *29.*

human ecosystems must either identify or assume the comprehensive effects these conditions have on human ecological complexity. For example, gravitational forces (which "order and orient," in Smil's fine phrase) create tidal cycles key to human activities from sustenance of coastal economies to art, craft, and myth-making. Base conditions limit, constrain, influence, and sometimes direct many human ecosystem processes. Hence, base conditions undergird all realistic efforts to understand the structure and dynamics of human ecosystems.

3

In the HEM (figure 2.1), a key *element* of the model is a set of critical resources. These resources are required in order to provide the social system with necessary supplies. These resources are made up of three *components*, each with a set of variables: *biophysical resources* (energy, air, water, land, flora, fauna, nutrients, and materials), *socioeconomic resources* (information, population, labor, capital, and technology), and *cultural resources* (organization, beliefs, art, crafts, and myth). These critical resources keep the human ecosystem functioning; their flow and distribution are critical to ecosystem coherence and sustainability over time.

Some critical resources may be indigenous to the local area and used locally or exported. Others may be imported from adjacent or faraway locales. For example, urban sources of investment capital and national media sources of information generated in Kuala Lumpur are integral parts of Malaysian rural human ecosystems. Soil, land, and nutrient resources in Costa Rica may provide inputs of coffee to harried and addicted American commuters; such "ghost acreage" is an arguable component of both the human ecosystems importing and exporting such supplies.[5]

The flow and use of these critical resources are regulated by the other key element of the human ecosystem, the *social system,* defined as the set of general social structures, patterns, and processes that guide much of human behavior. The social system is composed of three components. The first component is a set of *social institutions,* defined as collective solutions to universal social challenges or needs. For example, the collective challenge of maintaining human health leads to medical institutions, practices, and practitioners, which can range from traditional shamans and *vodouns* to modern hospital systems, western-trained physicians, rural health cooperatives, and preventive care. Other social institutions deal with universal challenges such as the need for kinship (family), health (medicine), sustenance (agriculture and resource management), justice (which can be expressed by law), faith (religion), commerce (business and industry), education (schools), leisure (recreation), governance (politics), and security (trust).

The second component of the social system is a set of *cycles*, the temporal patterns for allocating human activity. Time is both a fixed resource as well as a key organizing tool for human behavior. Some cycles may be physiological, such as diurnal patterns of sleep and rest. Others may be institutional, such as permitted hunting seasons or trading hours. Still others may be specific to the individuals or groups of individuals, such as work schedules that vary by occupation. Environmental cycles range from drought cycles to hurricane seasons to long-term climatic change. We note that environmental cycles have significant social components, both in their causes and consequences—global climate change being the most pervasive, anthropogenically driven, and powerful. Timing cycles significantly influence the distribution of critical resources. An example is the set of collective rhythms within a community or culture that organize its calendar, festivals, harvests, fishing seasons, war fighting, *carnivale*, business days, and so forth.

The third component is the *social order*, which is a set of cultural patterns organizing interactions among persons and groups. The social order includes three key mechanisms for ordering behavior: personal identities (age, gender, class, caste, and clan), norms (both formal and informal rules for behaving), and hierarchies. There are several kinds of hierarchy: Wealth, power, status, knowledge, and territory are among the most critical. These hierarchies work in combination with each other and with the other components of social order. Hence, strong predictions about interactions are possible when one can identify the age, gender, status, and power of persons or groups, and such expectations allow the social system to function.

The social order—individually, collectively, and in relationship to social institutions and timing cycles—provides high predictability in much of human behavior. Taken together, social institutions, cycles, and the social order constitute the social system. Combined with the flow of critical resources, these variables create the basic structure of the human ecosystem described in figure 2.1.

Finally, within the basic structure of human ecosystems, there are several key *flows*. It is these flows that create the dynamics of human ecosystems. Six seem essential. The first is the flow of individuals, which may be flora and/or fauna of varying species, including *Homo sapiens*. The second is energy, in a variety of forms from mechanical to heat energy. The third flow is nutrients, including food. A fourth key flow is materials. In this usage, materials include biophysical resources such as water and wood and man-made materials such as concrete and cocaine. The fifth is the flow of information, which can range from genetic to cultural and has been significantly affected by technology and the use of "social media" in recent years. The sixth flow is the flow of capital, in money form or its bartering equivalents.[6]

4

The HEM we have presented includes six major *components*, in the two major *elements* of critical resources and the social system. The six components are each made up of several *variables*. There are also six key flows that are variables in the model. While not simple or amenable to acronym, the complexity is tractable.

Consider the following metaphors. A Cessna Piper Cub II prop-engine airplane has approximately 50 warning lights and performance indicators; a 747 commercial jet aircraft has approximately 500; the U.S. Space Shuttle has approximately 5,000 of these indicator instruments. Our model of human ecosystems operates at the rough magnitude of Cessna complexity—demanding, but not overly so. (Of course, to us, the relationships within and between human ecosystems are infinitely more interesting than the mechanical systems of the Cessna.) Similarly, consider the study of relationships in a traditional "nuclear" family of mother, father, and offspring versus the complexity of the relationships of extended families with not only the traditional grandparents, aunts, uncles, and cousins but today's blended families with multiple step-relatives, half-sisters, and so on; or versus the complex families in many developing countries where entire villages are considered "families" by those living in the village.

The key is that this model complexity is driven by observation of real-world human ecosystems. Each of the components and variables plays an important role in our envisioning of human ecosystems. Careful attention to their individual and collective description can both explicate the model and demonstrate the exuberant nuances by which human ecosystems vary and function. Take energy flow, for example—ubiquitous among human ecosystems, yet with enormous range of expression. The wood-cutting preferences and habits of New Englanders versus Idahoans versus Nepali villagers reflect *both* regularity *and* diversity.

In the realized world—and in individual human ecosystems—the two primary elements of critical resources and the social system are expressed in a continuum of constructions. For example, qualities such as preferred tastes (sweet vs. sour, hot vs. cool) are at once chemical reactions and cultural responses that affect what materials and nutrients are acceptable for sustenance within a specific human ecosystem. The growth of woody biomass in forests is dominantly a biophysical response to the base conditions of sun, water, soil, and so on. But social system components such as social institutions involving variables related to kinship, sustenance, and commerce are affected by forest management decisions to harvest wood products or use prescribed fire to reduce fuel loads in a forest. Such decisions can have powerful effects. Other variables, such as leisure (recreation or play), are primarily defined as part of the social system. However, biophysical resources with the variables land, water, flora, and fauna may define wilderness landscapes and can have powerful effects on leisure. A continuum does not necessitate confusion or ambivalence: The gospel singing of Sam Cooke and the Soul Stirrers is at once the physics of air movement and the emotion of the choir; baseball is both Newtonian movement and the stratagems of teamwork. Yet in such cases we can easily place gospel music and a baseball team as social factors.

In the next three chapters, we describe the evolution of the model over nearly a half-century of our collective work.

T·H·R·E·E

Lessons and Legacies

1

The Human Ecosystem Model outlined in chapter 2 consolidates and restructures some of the systematic thought and action on human-nature transactions. Specifically, it outlines a framework for decisions that are effective, efficient, and equitable in sustaining our human and biophysical ecology. Not surprisingly, professionals and concerned citizens have generated many differing perspectives on the nature of the problems, the theories about cause and consequence, and the methods for resolving the perceived issues. The HEM seeks to fit this scientific and experiential information into a more manageable and cumulative organizational frame. This frame reduces much of the information "noise" to some key variables and relations that guide crucial questions about patterns and processes of the human ecosystem.

Present-day natural resource planners and managers may assume that they have unique environmental and natural resource challenges when compared to the past. However, one lesson learned in the early development of the HEM is that though the names and topics may change over time, the problems and solutions share very similar characteristics and replicable practices. That is, these can be *ex post facto* mini-social experiments. Ideas about "timber famine" and unstable rural communities needing "sustained yield forestry" and cries that the establishment of wilderness areas and other protection strategies will destroy jobs—these are echoes from the twentieth century heard in the twenty-first century Silver Valley of Idaho. Workers at the Superfund sites of the U.S. Environmental Protection Agency (EPA) in 2014 are greeted by local bumper stickers that read, "Hey Hey EPA, How Many Jobs Will You Kill Today?" Debates over fracking to gain oil and gas resources and its long-term impact on local communities and debates over global climate change and the impact it may have on human communities could as well be debates about wilderness, timber harvesting, mining, and the impact on salmon reproduction from dams and water pollution by paper mills in the 1940s and 1950s.

There is high utility in capturing lessons from this continuity of natural resource issues and applying the knowledge and experience gained to guide decisions for present and future environmental issues. Building on these lessons, we created a conceptual framework that defines the human ecosystem and serves as a foundation for the unified logic of inquiry that is the HEM. To understand the utility of the HEM, it is important to understand the interplay between a conceptual framework and a conceptual model.

A conceptual framework is a means for thinking systematically about our intentions and understanding of a given object or perception, while a conceptual model describes the structure and function of the actual outcome sought. For example, in 1973 Burch and his partner, along with a second-year Yale architecture student, Don Raney, built a house to minimize its impact on the environment and to provide passive solar energy opportunities. To do this, the architectural design did not just include the house but also incorporated elements and the manipulation of those elements from the surrounding biophysical environment. Burch describes building the house as follows:

> One thing you learn when you build a house is how important the foundation is. If it is not done well then the whole enterprise is at risk. So for the house we moved from our concepts that defined what we wanted—light, view, sunlight, shade, and coherence with the physical reality of the site we

had chosen for placing the house on its landscape. This was our conceptual framework.

Then we needed a means for seeing how the house would look inside and outside and its coherence with the realities of the building site. This was the model for placing the windows, rooms, plumbing, and so on. It instructed us how the vision could be implemented. It kept our focus during construction when we needed to adapt to the reality of the landscape characteristics and the permission environment of the electrical, plumbing, water, sewage and zoning codes.[1]

The humble act of building one's home is a microcosm of the larger scale of issues involved in the research, policy, planning, and management actions necessary for achieving sustainable human ecosystems. The energy crisis in the 1970s challenged physical and social scientists to come up with some usable answers for the causes and consequences of this challenge and what might be done to resolve, mitigate, or adapt to the problem. Clearly we needed some usable tools to frame the necessary and sufficient means for better understanding the nature of transactions between humans and their biophysical world.

Our start was development of the human ecosystem conceptual framework. To develop this framework, we looked for lessons from past work and found a very large base of knowledge. The basic sources range from rural sociology case studies to theories and reports on community development projects and applications. Several academic disciplines have edge interests such as the geography of local places, social psychological studies on the meaning of nature for individuals, and anthropological studies on the role of nature in culture. Diaries of travels by naturalists such as William Bartram, John J. Audubon, John D. Olmsted, and many other pioneers have described and sketched the embedded reality of humans within the natural landscape. This literature expressed a universality of behavioral reality that crosses time and spatial boundaries. This base of documented empirical work ensures confidence that when using the model, our decisions are not made in isolation and uniqueness but rather grow out of prior experiments on what works, what does not, and why. These questions express a trend of events surrounding similar issues in different times and places that collectively identify probabilities of likely behavioral responses to such issues in the future.

2

Walter Firey, in a classic study published in 1960, used empirical cases on land use practices and their constraints to understand the dilemmas and the opportunities for such decisions and to provide a theory for better environmental policy and planning. His theory provides a logistical base for thinking about complex environmental decisions. There is the ecological element that shapes what is possible and not possible, the economic element that shapes what is gainful and not gainful, and the ethnographic element that shapes what is adoptable and not adoptable. The policy-maker or planner must juggle these sometimes irresistible and incompatible elements in decisions. The concepts used provide a matrix for thinking about the possibilities and constraints that shape these three interacting elements of the human ecosystem. As Firey notes:

> The resource planner needs more than a theory of how resource users *actually* behave; he needs some "as if" theories, some fictional theories, as it were of how people ought to behave if certain abstract limiting conditions are to be *approached* (even though they can never be reached). Such theories provide the planner with some generalized reference point against which he can plot the relative position which a given resource occupies... The theories of ecology, ethnology, and economics offer just such reference points . . . though they can never be simultaneously achieved in real life, [they] nevertheless serve as ideal standards from which a resource system departs at the cost of predictable consequences. The wise planner will use them as canons of what ought to be but can never be. He will turn to a more realistic theory to discover what *can* be. For in his own practical way he knows that man is both a destroyer and a creator of natural resources and is unlikely to ever be otherwise.[2]

The famous wager between Julian Simon (an economist) and Paul Ehrlich (a biologist) over the idea that an expanding human population would place unsustainable demand on limited resources (with its echoes of the Faustian wager) is a classic example of what Firey is examining.[3] Ehrlich concentrates on the destroyer-of-resources characteristic, and Simon concentrates on the creator-of-resources characteristic. Firey emphasizes that the correct mix for such decisions is not to be found at either extreme and must consider the influence of cultural factors as well. And it is the interacting of these elements, not their separation, that may lead to more sustainable decisions.[4]

The HEM helps us to navigate through these complexities. It does not give a final or absolute solution for environmental problems. Rather, it gives a focus for asking critical questions about a given resource issue and a means for identifying an array of choices and the relative and likely consequences of implementing one or another of them. The HEM works within the complex reality that Firey analyzes.

The reader may wonder: Why would a theory about the processes that shaped historical land use patterns be of use for environmental decisions in the twenty-first century? Consider that the acid rain that kills lakes and ponds in northern New York and New England depends on land use decisions made in Ohio and Indiana. The acid rain that kills fish in lakes and ponds in Sweden and Norway reflects land use decisions made in Germany and France. The air pollution in great urban centers of the world depends on the conversion of forests to farms and abandoned farms to residential sprawl, a process that is shaped by decisions that encourage such development, including decisions about zoning, mortgages, highways, and perceptions of the good life. In short, all environmental issues begin and end in decisions made about land use. Firey's theory is a good place to start a decision model for better environmental decisions.

The second influential theory comes from the Hubbard Brook ecosystem studies led by ecologists Frederick H. Bormann and Gene Likens. They insist they are developing a "biomass accumulation model," and we will look at that model in a following section.[5] However, in social science terms, they also provided a midrange theory about the structure and function of ecosystems, in particular the mechanisms and cycles affecting watershed ecosystems. Burch was introduced to the Hubbard Brook studies in the early 1970s as a friend and colleague of Bormann at Yale. At the time, Burch was the chair of two social science groups—rural sociology (natural resources) and sociology (environment)—who thought they had much to contribute to the conversation. In listening to papers and discussions at these meetings, it seemed that the lessons from Hubbard Brook could liberate these inspired scholars from the stringent frame of traditional human ecology. Indeed, they could rejoin the life sciences and build a more effective discipline by adopting much of the theory and approaches from this robust version of biological ecology.

A special lesson from Bormann and Likens (similar to Firey) was their humility. They note at the outset the challenges of working at the system level of ecology. In the introduction they write, "We have no grand computerized model where all the animate and inanimate components and processes of the dynamic ecosystem are elegantly linked and where the details of the interactions can be spilled forth by a conversation with the computer. Indeed no such model exists for any ecosystem . . . our major integration tool is our desire to integrate rather than dissect . . . at this stage in the evolution of ecosystem science, a carefully defined and documented case history would provide the best vehicle for generating principles regarding structure and function of an ecosystem."[6]

The Hubbard Brook research looks at the biogeochemical cycles in a northern hardwood forest under different experimental treatments. They measure the role of biotic regulation of the system under different conditions of perturbation and note:

> Ecosystem development may be considered a battle between the forces of negentropy and entropy or between development of the ecosystem organization and its diminishment . . . some aspects of the control of destabilizing forces in an ecosystem are analogous to a controlled nuclear reaction, in which the enormous energy of the atom is released in small useable amounts rather than in one big bang. Movement of water through an ecosystem, for example, bears many

similarities to a controlled chain reaction. Precipitation is
first intercepted by the canopy, then by litter on the ground.
It is channeled by soil structure through the ground rather
than over the ground, and before stream flow from the ecosystem can occur, hydrologic storage capacity must be satisfied. Storage capacity is continually made available by evapotranspiration. Water enters the forested ecosystem with the
potential of a lion and, most often, leaves meek as a mouse,
with much of its potential energy lost in small frictional increments or simply by conversion of the liquid to vapor by
evapotranspiration.[7]

This theory of cumulative stacking of relationships between elements is very much a guiding force in the early development of the HEM, with its interest in second- and third-order causes. Frank Golley, in a summary analysis of the ecosystem concept being applied by biological ecologists, emphasized the uniqueness and utility of connecting system elements rather than concentrating on a few functions of specific elements.[8]

The Hubbard Brook studies and their conceptual framework are highly useful in structuring similar questions that join interests of environmental social scientists.[9] Further, the Hubbard Brook studies encouraged the authors to pare down the array of choices as to the critical minimum set of variables and cycles of connection for the HEM. This meant a more efficient and directed means for ensuring more effective natural resource decisions. For example, consider the decisions needed to sort out the consequences from an ever-expanding menu of energy choices: wind, solar, geothermal, nuclear, coal, hydro, conservation, petro, and biomass. We see a system becoming ever more complex even at the seemingly simple level of technological innovation. Humans have been using wind systems for many generations to do work such as pumping water and powering ships. Here is the creator side of resource development. However, contemporary wind systems designed to produce electricity require larger numbers of large windmills often located in areas with small numbers of human residents but within attractive landscapes. Thus, a seemingly benign technology has large impacts on aesthetic settings, enrages nearby residents, kills migrating birds, and requires toxic storage units when the wind does not blow.

Those who engage in research, policy, planning, and management of complex systems must be aware that there are wide-spreading and often unknown dimensions of "collateral damage" from establishing these systems. Here, the usual strategy of traditional science research, in which we try to reduce the system to separate mechanisms, may be the problem rather than the solution. Complex systems can only be understood by working through their complexity rather than trying to avoid it.[10] This requirement means that there will be a significant shift in the theories and models we use to understand and to work with such systems and a challenge to traditional patterns of thought and action. The Hubbard Brook influence on the design of the early HEM gives legitimacy for resource professionals and local communities to adapt to these new realities rather than default to traditional practices.

We will round out our examination of the representative theories that influenced the architecture of the HEM with a look at two social science midrange theories—Diffusion Theory and Symbolic Interaction Theory. We highlight these two theories because they give substance to many of the connections necessary for the functioning of the HEM. The first theory systematically analyzes the conditions that lead to rejection or adoption of natural resource innovations in technology or policy or social values. The second theory highlights the behavior of individuals as they operate within the social roles they adopt or are assigned and how these interactions are likely to play out in terms of conservation or depletion of natural resources.[11]

If we treat theory as "a set of logically interrelated assumptions from which empirically testable hypotheses are derived," then diffusion theory may seem more like a model than a theory.[12] Anthropologists use it to see how cultural traits are adopted (or not) in exchanges from one cultural group to another. Many rural sociologists and economists use it as way to explain how technologies and new products gain traction in desired markets or with targeted populations. Rural sociologists are interested in processes of adoption or rejection of technologies that increase the agricultural productivity and income of farmers. Economists are interested in rates of adoption by certain consumer populations for specific new products such as a radio or a smartphone.

Global economies, population growth, and the increase in non-point source pollution suggest the need for engaging local persons and communities to become stewards of local landscapes. In this way, ecosystem management becomes a responsibility of all those whose wise use strategies accumulate to provide a more sustainable world. It is a "bottom-up" rather than a "top-down" management strategy. Here, knowledge about the processes involved in the adoption of innovative ideas of ecological stewardship would be crucial.

Jared Diamond's analysis of how some societies became rich and others did not is based on diffusion theory in combination with geographic determinants to explain his observations. In looking at the diffusion of writing and agricultural innovations, he emphasizes the importance of diffusion from one place and culture to another. He notes, "Corn diffused from Mexico to the Andes and, more slowly from Mexico to the Mississippi Valley. But . . . the north-south axes and ecological barriers within Africa and the Americas retarded the diffusion of crops and domestic animals. The history of writing illustrates strikingly the similar ways in which geography and ecology influenced the spread of human inventions."[13]

Those researchers looking at diffusion processes over geographic scales smaller than those considered by Diamond tend to look at more structured processes based on characteristics of the adopters and interactions between the innovations being adopted and the reference groups that influence the behavior of the potential adopters or non-adopters. Of particular interest is the timing of the adoption—early to late, which follows an S-shaped adopter curve—a few venturesome folks at the start and a cumulative growth rate in the middle and a tailing off at the end.

The HEM gives a means for understanding the nature of the innovation needed—say, getting households to use less or no fertilizer on their lawns to reduce nonpoint source pollution of watersheds. Then there are categories for asking questions about interactions among the characteristics of the innovation (very green grass), the potential adoptee (a neighborhood household), the social network/reference group (friends and neighbors), the adoption leader (a sports celebrity), the source of information (television ads), the trialability of the treatment (the neighbor's yard has had the treatment for the past three years), and the means or visibility to monitor the impact of the adopted practice (comparison with other neighborhoods). The theory directs the HEM user to pick those variables most likely to influence behavior of the persons within the aggregate pool of potential adoptees.

The second theory that significantly influenced the early development of the HEM comes from social psychology—symbolic interaction. Arnold Rose provided one of the best short descriptions of its origins. He notes, "Symbolic Interactionist theory . . . had its American origin around the turn of the [twentieth] century in the writings of C. H. Cooley, John Dewey, J. M. Baldwin, William Isaac Thomas and others. Much of the theory had an independent origin in Germany in the writings of Georg Simmel and Max Weber. Its most comprehensive formulation . . . is the posthumously published volume by George Herbert Mead, *Mind, Self, and Society* (1934)."[14]

For the purposes of the HEM, the unique vision of Mead is that we exist in a symbolic as well as a material world; that is, communication is a critical element of the human condition.[15] We share with nonhuman animals the use of signs and gestures to communicate with our conspecifics. Symbolic communication seems to be a unique feature of human communication. While signs point, symbols re-present objects, persons, events, and ideas in such a way that the external world is filtered through the lens given to us by those who taught us our language and those with whom we regularly exchange messages. This lens of meaning is loaded with values shared among a "generalized other," the communication networks that socialize a person in terms of social roles and the norms expected of role occupants. A child touches a hot stove and burns his or her finger, while a parent says "hot, hot, hot," and the meaning of the danger inherent in the stove is summed up in the word *hot*. And that word now directs the future behavior of the child. This theory identifies the mechanisms of meaning and their ability to direct the behavior of the individual self. So words can sum the meaning of abstractions like "liberty" or "justice" or "beauty." Those meanings come from one's primary reference groups, which reinforce their selected memory of the word's meaning.

Symbolic interaction theory was an essential driver in the design of the HEM, as the canopy of environmental meaning was determined by communication strands sustained by a person's critical reference groups. Stream pollution had several sets of meaning as one shifted among shared linguistic communities. These might be professional groups, status groups, or kinship groups that drive the behavior of an individual for given situations. In a park, we have a community of shrubs. For some persons, these shrubs are invasive weeds; for others, these shrubs are good places to stash drugs; for others, these are objects of natural beauty; for others, these shrubs mark the boundary between "my place" and "your place." We could go on. The reader can surmise that much conflict over natural resource matters occurs when people use the same word, for instance, *conservation,* but have many different meanings attached to that single word. This theory unlocks the utility of many of the concepts used in the HEM as it tracks motives for social meanings assigned to environmental events like global climate change or the evolution of lifeforms.

Symbolic interaction theory is a theoretical tool that helps environmental innovators to transcend their "trained incapacities" and respond to a wider environmental context. The following two examples may clarify the utility of reaching out to others whose meaning sets about *conservation* differ from that held by the innovator.

In the early twentieth century, Bernard Fernow was establishing a forestry field station in the western Adirondacks for Cornell University. He followed traditional German forestry practices, which were to maximize biomass of commercial forests. He clear-cut the native species and began to plant "fast-growing" European species of trees. However, he overlooked the social environment where he was demonstrating the "correct" forest practices. This environment was populated by wealthy, seasonally-resident New York families whose notion of an ideal forest was very different than that of Fernow's. The challenge by the wealthy owners of "camps" ensured that the infant forestry program at Cornell was dissolved, and in 1901 Gifford Pinchot started a forestry program at Yale University. An HEM-guided Fernow could have fit his forest treatment (flora and fauna) into cultural resources (beliefs) and considered the normative drivers of the human ecosystem in the Adirondacks such

Landis's work is important for a variety of reasons: It compares communities that share certain biophysical and economic realities yet vary in adaptability, giving it characteristics of experimental testing; it takes the importance of natural resources of minerals, forests, and water as central factors in shaping social life; and it gives attention to a variety of cyclical factors from demographic to biophysical as dynamic and likely overlapping attributes of cultural adaptation. As he notes in his preface:

> Few industries that establish community life are more short lived than mining. The short life cycle offers an unusual opportunity to bring the historical processes of social and cultural change of a community within the scope of a special study. The short span of time involved permits an accurate observation of the incidence of change; the speed of development permits observation of the processes of growth and maturity; the homogeneity of the economic and industrial base permits an observation little complicated by a multiplicity of distinct social classes and social standards. Under such circumstances one can be fairly sure of what is going on and why. Even though the study deals with a single area, the processes and relationships revealed no doubt characterize a larger universe.[19]

His methods seek reliability, validity, predictability, and universality. He clearly states his assumptions and hypotheses. He looks at the same Mesabi landscape, where different cultural values have perceived very different core resources. Native people saw it as a hunting ground; for the lumberman, "It was a country to exploit and to abandon"; miners saw it for its iron ore resources and as a place to build communities. The miners were being joined by "tourists who saw it as a paradise of fish, lakes, wildlife, and scenery." The definition of a natural resource varies by the perceptions held by different user groups. The physical, of course, is important, but the nature of what kind of resource it is comes from how specific groups of people sharing a vision of the landscape can reconstruct the meaning of that landscape in many important and significant ways.

Landis looked at town records and census information to note the more than thirty diverse linguistic groups in the three towns, and he tracked the rise of the foreign population over time and how these variations influenced cultural adaptability of the community. From these and other municipal and federal data, he charted the cycles in community life—the pioneer period of growth; the middle period of the mining civilization, marked by conflict; the final period of decay and disintegration. He overlooked the possibility of new technologies and changes in market values that could get iron from the taconite by mining the spoil sites of the early mines. So decline happened, but re-growth also followed on the basis of market values and technological change.

There are two methodological lessons to be learned from Landis's work. First, do not be overconfident about your charts and graphs. In the Mesabi case, the role of technological change was not considered. The second important lesson is that examining a variety of overlapping cycles is a useful way of teasing out possible causal connections. In addition to charting community development from the pioneer startup to the community decline cycle, he had parallel culture cycles and biophysical cycles that drove the observed community patterns. Landis notes cultural variations but links these to the natural resource base that serves as the stage upon which many social stories can play. The early HEM version used his lessons to offer a way to link cycles and processes to critical variables and to one another. These cycles provide predictability in terms of changing demand and changing consequences. This is similar to the cycles used in the Hubbard Brook ecosystem studies that describe the flux of minerals and nutrients influencing the pattern and process observed in the forested ecosystem.

The HEM takes findings and approaches from these historical studies and turns them into a base for analyzing contemporary natural resource issues. One example is Landis's use of age cycles in human populations coupled with other cycles to help predict coming demands on human ecosystems. We have good evidence of this in the impact of the "boomer" generation on health systems and other cultural resources, in which the increase in the proportion of older persons creates unanticipated costs that are likely to reduce funds for other government activities like pollution monitoring or the maintenance of parklands. There is similar concern about how prevailing patterns in age cycles challenge political economies in Europe, China, and North America with higher proportions of elderly being supported by fewer young persons. Or consider the age-sex ratio in China, where the now-abandoned one child, one family policy meant that more boys than girls were permitted to survive infancy; when this generation reached adulthood there was an imbalance of mates, which put extreme strain on the social structure. The HEM would have suggested to the policy-makers that cultural knowledge (belief) and social norms would determine that boys were likely to be preferred over girls. Yet that likelihood was not part of the enforced policy, and twenty years later a gender imbalance occurred. These two cases demonstrate that demographic fluxes are empirical realities that can predict and anticipate coming challenges to the human ecosystem.

Landis in 1938 was exploring the environmental and cultural consequences in the congruence between biophysical resources and dramatic shifts in demographic demands on human ecosystem resources. The consequences of the policies and forces that set in motion the issues noted above could have been anticipated by a careful use of the HEM. Though today we may have better sources of data and more efficient tools for analysis, the underlying logic of Landis's approach remains a substantial foundation for the HEM design.

Our second methodological example comes from a pioneering effort by the U.S. Forest Service and a local institution, "The Montana Study," conducted by Kaufman at the University of Montana. This was a period when the U.S. Forest Service was being pushed from being primarily a custodial program to become the primary source of commercial timber. This change happened at a time when postwar demand for housing and therefore timber was rising at a very high rate, and the agency was a major economic force in many western forest–dependent communities. Forest managers sought a means to have these communities not follow the path of high growth and then decline, as happened in mine-dependent communities. They hoped that the sustained management unit strategy would ensure community stability. It was what Samuel P. Hays called the "gospel of efficiency" that was promoted by the progressive conservation movement. The idea was that professionals, through wise ("science-based") management, could sustain the consumption of natural resources and thus sustain human communities. The Kaufman study was a first exploration as to how the life cycle of forest use could be shaped to fit the life cycle patterns of the human community.[20]

The researchers were in a difficult position. The agency wanted one set of confirming results for their new management strategy. The local citizens were concerned that most of the timber would be sold to the largest timber mill. The researchers immersed themselves into the social life of the communities to gain a feel for the attitudes and hopes of the local people. They analyzed trends in timber cutting and income over time and between different groups and related them to trends in jobs, local business, and so on. They held structured interviews with representative groups and regular informal meetings and discussions. They used public documents and other records to provide a useful profile of the communities. Within this context they examined how residents reacted to the Forest Service and the sustained forest unit strategy and what the Forest Service would need to do to encourage public participation and to serve community stability goals.

The most interesting aspect of their study was their attention to the nature, types, and trends of the formal social organizations. The population for Lincoln County in 1944 was 6,542, a 17 percent decline from 1940, while the Libby-Troy communities were 68 percent of the total population of the county. However, for such a small population, there were a large number of social organizations with active participation by their members. There were seven religious organizations, five civic organizations, seventeen fraternal organizations, five farm and home organizations, and two recreational organizations. They reported the number of members in each organization, average attendance, meetings per month, and date organized. The researchers then captured the influence that such organizations have on community attitudes toward the agency and its proposed strategies for stabilizing the communities: Their data captured both individual and organizational responses.

The study reported: "From the standpoint of the community members interviewed, the basic issue with respect to forest policy is not so much one of technical forestry as one of social control and economic as wealth, power, and status that trump narrow technical fixes that upset belief in the "proper" aesthetics of a forest. At the least the HEM would have helped Fernow recognize that many of the residents were active in the creation of the Adirondack Park in 1892, which was then made a part of the State Constitution in 1894 with a "forever wild" command.

This forestry example suggests that environmental realities are larger than one narrow focus on a primary outcome. An updated and upturned 2014 version of Fernow's dilemma is the desire of a wealthy individual, Roxanne Quimby, to donate over eighty-seven thousand acres of land to create a national park in the far north of Maine adjoining a major state park protecting Katahdin Mountain. Though a Maine native, Quimby initially overlooked the political power of local people of the region, who like to cut timber, ride snowmobiles and all-terrain vehicles, hunt and trap wildlife; many opposed the park idea. The clashing values made progress difficult. Quimby's son Lucas St. Clair restarted the proposal by listening to and working with local folk to find compromise with their wishes and the realities of a federal environment agency.[16] As a result, more local support was created, and on August 24, 2016 (the 100th anniversary of the establishment of the National Park Service) President Obama declared the establishment of the Katahdin Woods and Waters National Monument.[17]

An HEM analysis at the outset might have better prepared the innovators for adapting to the social environment and made the eventual adoption a more likely and less controversial proposition.

3

Two classic studies of natural resource–dependent communities serve as examples of historic guides in the development of the HEM. The first is Paul Landis's 1938 study of three iron mining towns in Minnesota, and the second is Harold Kaufman and Lois Kaufman's 1946 study of three forest-dependent communities in western Montana.[18] These two studies and others like them are the true godparents of most modern environmental social science. Both made solid research accomplishments and, most importantly, committed to giving voice to the people in the communities they studied.

reward. The control question is not rapid liquidation of the forest resource versus sustained production but rather concerns the problem of protecting the public interest and distributing equitably the rewards from the forest."[21]

Interestingly, telling the story as the researchers heard it was not a message the agency wanted to hear. In a letter attached to the final report, A. G. Lindh, assistant regional forester, apologized for a two-year delay and cautiously accepted the report. Then he wrote, "I have one major criticism of the report by Dr. Kaufman. It is not intended to be a criticism of his work since I think he ably and honestly carried out the plan which was agreed upon. The criticism is this: By implication and by occasional statements throughout the report, it appears that the Forest Service has conducted its planning and the management of its timber in the Lincoln County area without due regard for the people of the several local communities. I believe that I am correct in interpreting the instructions and Nation-wide policies guiding the work of the Forest Service when I say that the human and community dependency side of timber management planning is receiving and will receive its full share of attention."[22]

The Kaufmans were pioneers. It was not until the late 1960s that non-economist social scientists such as the Kaufmans were hired as full-time employees in the Forest Service. These heirs of the Kaufmans systematically collected information to help the agency give "full share of attention" to the wishes of the people even when the findings did not conform to official policy. Learning from the Kaufmann's work, the HEM stockpiles enough data and variables that it can float official "inattention" up to a higher reality about how human ecosystems actually work.

4

We have considered four theories and several case examples that structured the intellectual base and approaches of the HEM in its early design phase. Obviously, there were other influences on our design. For example, the work by Gordon Conway and other agroecosystem analysts in the research and development field (R. Levins and J. H. Milsum) was crucial. Of equal importance was the work of geographers such as

Howard W. Odum and Harry Moore and regional planners such as Lewis Mumford.[23]

However, the most influential model was the work of Likens, Bormann, and their colleagues that seemed to sum up most of the concepts and approaches of the other ecosystem models. They described the utility of their model:

> An ecological system has a richly detailed budget of inputs and outputs of energy and matter. Because of the lack of precise information about these relationships and the internal functions that maintain the ecosystem, it is often difficult to assess the impact of human activities on the biosphere. As a result, land-use planners often cannot take into account or even foresee the full range of consequences a project may have. Without full information, the traditional practice in the management of land resources has been to emphasize strategies that maximize the output of some desirable product or service and give little or no thought to the secondary effects. As a result, one sees such ecological confusion as an all-out effort to increase food production while natural food chains become increasingly contaminated with pesticides and run off and seepage waters carry increasing burdens of pollutants from fertilizers and farm wastes. Forests are cut with inadequate perception of the effects on regional water supplies, wildlife, recreation, and esthetic values, and wetlands are converted to commercial uses with little concern over important hydrologic, biologic, esthetic, and commercial values lost in the conversion.[24]

They used a conceptual model of deciduous forest ecosystems that has guided extensive and detailed interdisciplinary research work for over fifty years. The model measures input-output fluxes in cycling of water and chemicals. It measures the input of energy, water, nutrients, and other materials and their output in terms of geological, meteorological, and biological vectors. They used the watershed as an ecosystem boundary, and within this were four compartments: atmospheric, organic, primary, and secondary minerals. There was also an available nutrient compartment. They used this framework to guide the empirical work and the analysis of their data. The reader may find it profitable to investigate their work and that of their colleagues in further depth. For our purposes, we wanted to bring the human factor into an ecological system by using their work as the format for the design of the HEM. The intent was to have the human ecosystem be a part of the biological reality rather than simply tacking it on as an unruly external variable.

The HEM grouping of boxes of variables and input-output cycles tried to replicate the framework of the Hubbard Brook studies. In the early stages of HEM development, with its "Handbook for Assessing Energy-Society Relations," we grouped the human ecosystem variables into three compartments—supply elements, organized social order, and the social cycles that drive the system.[25]

We supplemented this array of variables and compartments with different variables and theoretical approaches—time budgets, adoption patterns, life cycle and demographic patterns, community structure and process, regional ecology, social survey tactics, and social indicators. Of course many other midrange theories could be identified, but our task here is to highlight some of those with considerable influence on the HEM and let them stand for these other options so that the reader can see their utility in developing the model.

The Hubbard Brook studies provided a model for how to organize an interdisciplinary research study program that combined a variety of applications in experimental treatments. It had watersheds as a definable unit of analysis to delimit boundary issues. It had experimental treatments and indicators to measure the response to these treatments. It had real-life suggestions for matters of timber harvesting, acid rain, and the results of several generations of human action. Its analytic metrics were based on indicators of empirical patterns. It did not attempt to ignore complexity but rather used this empirical reality as leverage for better understanding. Our translation of the biophysical metaphor of the Hubbard Brook framework to one that systematically included human behavioral regularities helped bring social science back to its connections to the biological realities facing our species.

Further, by working in rural and urban developing areas we were able to follow *ex post facto* kinds of research. Where the Hubbard Brook studies dramatically removed vegetation from some watersheds and left others alone, we started with heavily impacted, impermeable landscapes and then in some areas rehabilitated them, and in others we had relatively coherent systems to compare with the impacted sites. So a similar gradient of impacts could be re-traced in urbanized ecosystems. Our conceptual translation permitted a common language for dialogue with biophysical researchers.

5

The "energy crisis" of the 1970s, along with new government concern about water and air pollution, wilderness protection, timber cutting, and mining practices on the public lands, emerged with other changes in public awareness about the political "inequities" in many traditional resource practices. These issues gave a necessary opportunity for social scientists to work with biophysical management agencies like the U.S. Forest Service, the National Park Service, the Bureau of Land Management, the U.S. Fish and Wildlife Service and the Army Corps of Engineers, and later the Environmental Protection Agency. The challenge facing federal public land agencies was to participate in developing holistic, interdisciplinary models, methods, data, and insights. The leaders of these agencies were seeking guidance on how to respond to changed perceptions of traditional public policy regarding the use, demand, distribution, types, equity in distribution of burdens, and benefits from natural resource systems such as energy production, old growth timber management, and wilderness recreation.

The stage was set for application of the HEM. In 1977, Burch was invited to be part of the "Inexhaustible Energy Resources Study" directed by Bennett Miller of the new Federal Energy Department. In this meeting, many of the biophysical colleagues seemed unaware of the gains made by social science since Herbert Spencer's nineteenth-century Social Darwinian evolutionary theories. Indeed, there was a good deal of solid empirical information and systematic theories from functionalists, diffusion theorists, symbolic interaction, and other midrange explanations. Further, this knowledge could be assembled into a handbook (see the earlier discussion) that could guide policy decisions on energy and other natural resource issues. The impetus was for our team to put together a conceptual framework, provide examples of indicator measures that matched our concepts, create "workable" diagnostic means that linked critical elements of the system, and help "engineers, biologists, physical scientists, land managers, legislators, and other decision makers to direct and interpret work by non-economic social 'scientists.'" The HEM emerged as a social science contribution and application to the major environmental issue of that time—the decline of cheap and plentiful petroleum.

Our interest was a framework for understanding what each energy system requires from its social organization to carry out each of its several phases—extraction, production, processing, delivery, end use, and waste disposal. These organizational patterns involve different *social roles,* norms, and social institutions. That is, there is the need for social roles such as nuclear engineers, meter readers, or coal brokers, depending on the energy system. Further, there are rules as to how a person in the role of nuclear engineer should perform, as well as to how electricity rates and coal prices should be set. Finally, there are *social institutions,* which make routine the variable demands made on particular clusters of activities, roles, and norms. For example, electricity rates are usually set by state public utilities commissions with fixed rules and hearing procedures that involve social roles such as economists, corporate lawyers, consumer representatives, technical witnesses, and so on. The process by which rates are set, with its prescribed roles, rules, and timing, is *institutionalized* so that conflict proceeds in an orderly manner. Similarly, the price of coal is controlled by the social institutions of commodity markets, while the construction of nuclear power plants is controlled by the social institutions of a federal agency. Analysis of any particular energy system must consider the range, types, and order of the roles, norms, and institutions that impose a high degree of predictability on behavioral outcomes.

For example, a woodlot home-heating fuel system is labor-intensive and involves a family division of labor at various stages of the production-consumption cycle. The assignment of the various production roles is likely to be primarily based on certain age and gender characteristics. In rural Nepal, women are the primary gatherers and users of wood. In the rural United States, the primary gatherer and user of wood fuel is likely to be an adult male.

Electricity produced by a coal-fired plant has a different pattern of organization, and production roles are likely to be assigned on the basis of function, expertise, and technical skills, whereas a nuclear energy system is likely to have a higher proportion of its production roles assigned on the basis of formal technical training. For the household, regardless of the source of electricity, the social organization for consumption will remain fairly similar.

Our conceptual framework attempts to remain at the level of analysis where social roles, norms, and institutions are constant elements over time and space. However, as in the energy examples above, the specific characteristics vary by the nature of the particular human ecosystem function or process. It does this in two ways. It treats the social system as part of the biophysical environment rather than compartmentalizing the potential consequences into economic, biological, social, and physical responses. Second, it attempts to combine the approach of the ecologist, epidemiologist, biosocial anthropologist, economist, and sociologist into a single conceptual system. Figure 3.1 shows the interrelated variables that create a balanced system in dynamic equilibrium, organized through social cycles that, in turn, place new demands on habitat flows. During times of change, perturbations occur with either the habitat flows or within the social order, as diagrammed in figure 3.2.

Figure 3.2 gives a range of possible outcomes for the decision-maker to consider. The change in the amount or the type of energy converted from human labor to the use of a horse or a restriction in an available set of nutrients for members of a community (such as the absence of salt) could be read in how it affects certain elements in the social status system or myth cycles. The tool in this case is the framework that calls up the likely factors driving stability or change in a given system. The person using the HEM has a tool for asking better questions about what is going on in and between the system elements and its environment.

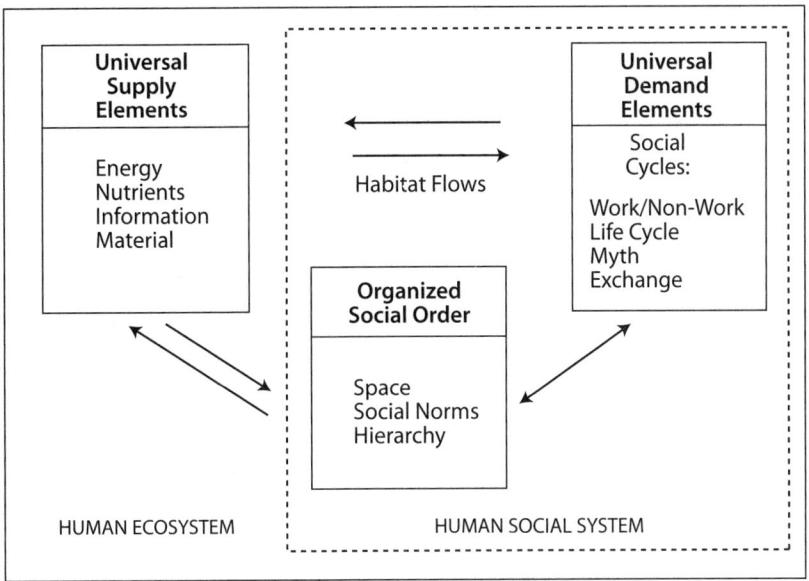

Figure 3.1. Generalized conceptual diagram of the social system within the human ecosystem

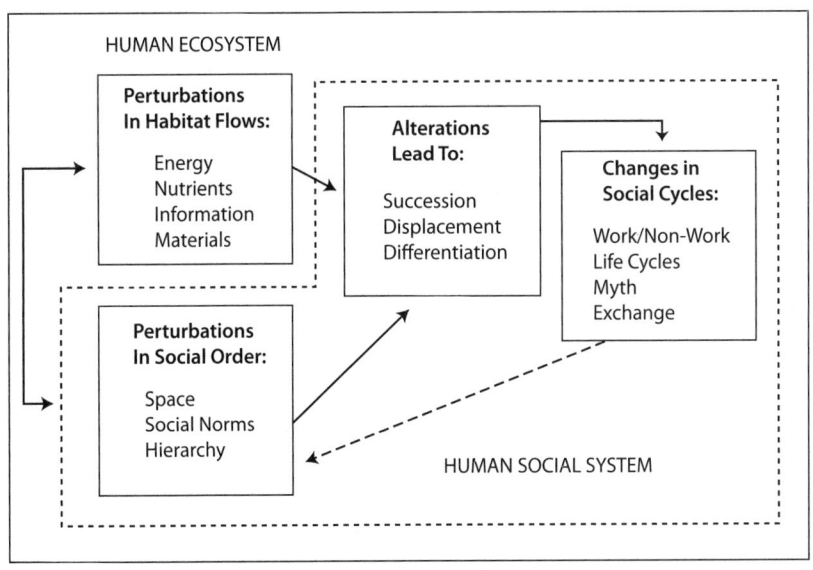

Figure 3.2. A conceptual diagram of the social system within the human ecosystem: a dynamic equilibrium model

A good example for putting into play the conceptual model is the story of the re-introduction of the horse to the Americas by the Spanish. This innovation led to a dramatic change in the life of the Plains tribes and had a large impact on nonhuman ecosystems for those tribes who adopted the horse.[26] Here was a major perturbation for the Plains Indians. Suddenly an energy converter many times more effective than the dog became available. The tribes with the earliest access to the horse greatly expanded their range and ultimately displaced many of the more sedentary cultures. The social cycles were markedly affected. The food surplus led to an increased population and a lowered median age. The patterns of work and nonwork were determined by the migratory patterns of the bison rather than the growing seasons of cereal and tuber crops. There was greater attention to ornamentation, religious feasts, and personal wealth. The commuting distances for work were greatly increased. The bison and horse began to take prominence in the myths, rather than the sun and the rain. The exchange cycles expanded to wider distances with more "luxury" products and increasing dependence on alien suppliers. These changes altered attitudes toward the "appropriate" size of a territory and the "appropriate" behavior of the priesthood and warrior classes and moved from more egalitarian patterns of authority to the more stratified patterns of large-scale warrior tribes.

There was cultural succession in that new patterns of land use and property rights contrasted with those of the more sedentary and smaller bands. The characteristics of the horse and its ownership were more important than landed property rights. Further, in the traditional small bands the primary division of labor was based on age and gender; this division of labor now expanded to a wider range of social positions and roles. The wider mobility of the horse converted the Plains to a vast open range and replaced the more limited hunting-gathering economy centered around the dog. The displacement of the sedentary peoples near the buffalo range by nomadic mounted warriors brought about social disruptions. There was increased occupational differentiation as various classes of warriors, chiefs, priests, captives, and slaves emerged.

6

The HEM provides an intellectual venue for linking several kinds of methodological perspective. The benefit of this consolidation becomes apparent when we consider just three different perspectives—biology, epidemiology, and environmental anthropology—and how they can inform understanding of the trends and consequences in human ecosystems.

In their discussions of the potential impacts on an ecosystem, biologists often talk in terms of the disruptions to nutrient and energy flows, of the influence on the stability or instability of certain information patterns as genetically derived, of altered population dynamics for certain species, and of the alteration in patterns of hierarchy within a given community or species. In short, the biologist looks at the survival potentials of certain species, groups of species, or entire ecosystems and the mechanisms available for short-run and long-run adaptability. "Bad" actions, in biological terms, are actions that irrevocably destroy the habitat of particular species of flora and fauna. The loss of a particular habitat and its associated lifeforms does not mean the end of evolutionary processes or even the end of the earth. Such losses are part of evolution. However, biologists might argue that accelerating and incremental loss can accumulate in such a way that ultimately human values, human societies, and even human survival will be significantly affected.

Epidemiologists and public health professionals identify specific health effects that are "caused" by certain environmental perturbations. By and large, the statistical associations they find between specific disease episodes and specific alterations in air or water quality are based on aggregated data. Thus, certain levels of air pollution tend to increase mortality rates for certain susceptible persons—the elderly, the young, or persons already suffering from mild bronchial problems. Their studies permit broad statistical predictions of the nature that a particular industry with a known process and a known type of emission system is likely to produce an increase of so many grams per square meter of particular pollutants, which will result in certain changes in morbidity and mortality rates for specific segments of particular populations.

Some social anthropologists combine their interest in cultural influences and diffusion of technologies or ideas that lead to measurable changes in social organization, which in turn have altered nutritional patterns in such a manner that the new organizational and hierarchical system becomes self-perpetuating through biological factors—such as the lead used in Pompeian wine containers resulting in sustained damage to the brain, thus limiting effective participation of certain social classes. That is, under the impact of the change, certain social strata become markedly better off and other social strata become markedly worse off, and these differential patterns become biologically sustained. In this sense, a process difficult to reverse has been established, and the alteration in the hierarchical structure suggests the decay of one community form and the emergence of another. In these biosocial anthropological studies, a good deal more complexity has been added to the epidemiological approach, but the issue of survival remains central.

7

In this context, given some of the realities of applied science, our early HEM represented a radical approach. We sought to develop a model that provided a means for joining that which must be joined if we are to build an enduring compact between humanity, with its many diverse players and perceptions, and nonhuman nature, which has its own cast of supporting players and perceptions. This diversity of participants and the search for a unified approach was clearest in the work of a long-term seminar at Yale during the 1970s—"The City and its Environment." This was an intellectual mixing bowl where life was stirred into the HEM. It was led by Charles Walker, a senior professor of engineering, along with distinguished emeritus professors like the psychologist Leonard Doob, who were joined by graduate students (including Machlis) and faculty (including Burch) from a diversity of disciplines.[27] There was also some modest support from the Conoco Corporation, and sometimes their executives attended the discussions and presented private-sector perspectives. What was most clear is that all agreed that we could not make progress on understanding without genuine interdisciplinary efforts. However, this consensus could not be a reality for untenured junior professors whose discipline required full attention to the narrowed domains of their specialty.

Though not all could play in the unity dialogue, the facts of universality and uniqueness were a concordance of faith. To that end, the HEM is a useful script for the decision-maker, the student of this particular place in this particular time, and it can be the essential instrument for understanding the human system's many possibilities. The key to this effort is the decision-maker, as Firey reminded us, who has the flexibility of the moment, the experience of this "other," and the observational skill to make a studied judgment. The HEM provides a means to do a better, more comprehensive assessment of what is going on and what are the most likely, most effective, and most efficient ways to go about dealing with a given natural resource or environmental issue. In ecosystem studies and actions, the key to the effort are the trained observers who are aware of their responsibility in serving as the catalyst for the integrity of the decisions. As Timothy Allen and Thomas Hoekstra have argued, "The essential beauty of ecological material can be seen with remarkable clarity through the eyes of the manager. The naturalist and the preservationist do not have a monopoly on the joy that is to be had from being an ecologist in the woods. Management is a very esthetic matter. In the modern biosphere, human activity is part of the system in a new dynamic interplay. Our species has the next dance with nature, and it is the ecological managers who should be the dancing masters, and the orchestra leaders."[28]

8

The HEM draws on an intellectual and practical base derived from the work of many others in many disciplinary and practical efforts, and these ideas still guide the work. Much of the work clusters around issues such as the decision-making process dealing with energy issues and, more broadly, questions about how to have effective, efficient, and equitable policy, planning, and management actions for sustaining critical natural resources. Environmental impact assessments, mitigation studies, and models for making urban and wildland decisions and evaluating environmental resource policy, planning strategies, and management actions may all be based on the HEM. Other possible applications include designing interdisciplinary socioecological research studies or predicting social ecology considerations for disaster management and other risk assessments.

The HEM holds to this pattern of baseline understanding. It does this in two ways. It treats the social system as part of the same biophysical environment rather than compartmentalizing the potential consequences into separate bins of economic, biophysical, social, and ecological responses. Second, it attempts to combine the approach of the biological ecologist, epidemiologist, biosocial anthropologist, and social ecologist into a single conceptual frame. We continue this conversation with an expanded intellectual context for the current organization of the HEM, as it moves from framework to model.

F·O·U·R

The Ecosystem Concept in Biology

1

Intellectual genealogy, or tracing the historical path of a scientific concept, is an essential foundation and a valuable method. It can suggest worthy avenues of discovery and synthesis and illuminate potential routes of difficulty or failure. Isaac Newton's sense of debt ("If I have seen farther, it is by standing on the shoulders of giants") and Stephen Jay Gould's reminder that "structures" in science are emergent and collaborative recommend a brief history of the ecosystem concept. In addition, the historical chasm between biological ecologists and social scientists (closing now, albeit slowly) recommends reflection on how the ecosystem idea has developed in the natural and social sciences.

We now examine the genealogy of the HEM. This includes portions of ecosystem biology and the roots of human ecology. The initial and repeated efforts at taxonomy (identification of component parts) were followed by efforts to understand processes (identification and measurement of flows) as well as issues of agency, efficacy, and evolutionary change. All are foundations for understanding the structure and dynamics of human ecosystems.

2

Studies of the patterns and processes in ecosystems emphasize the diversity and complexity of the elements affecting these systems. Likens and Bormann note that "a vast number of variables, including biologic structure and diversity, geologic heterogeneity, climate and season, control the flux of both water and chemicals through ecosystems." Golley re-emphasized ecosystem complexity: "The ecosystem consists of co-evolved suites of organisms . . . there are keystone species that provide special environments for many other groups. There are also social organisms, such as ants, that form yet another pattern of organization. This means that the actual organization of an ecosystem is much more complex than the network model suggests. Indeed, the organization of a large city might be a better model than the systems model of textbooks, the links of which if very complicated look like a bowl of spaghetti."[1]

This complexity has generally enabled biologists to exclude human behavior from their models (Golley's comment on the city notwithstanding) and social scientists to remain largely at the level of metaphor when applying ecological concepts to human behavior.

There is irony in the historical evidence that early ecologists freely borrowed from the social sciences to construct key concepts. Henry C. Cowles described "plant societies." Arthur Tansley borrowed from Herbert Spencer to create his "organism-complex"; he left ecology to study psychology with Sigmund Freud. Anton Joseph Kerner reasoned from human communities to "plant-species communities" and Stephen Forbes to "communities of interest" between predator and prey. Frederic Clements was influenced by Spencer and sociologist Lester Ward. Both H. T. and E. P. Odum were influenced by their father, H. W. Odum, whose sociological study of the American South was prescient in human ecology.[2] Wassily Leontief borrowed his input-output analysis from economics and applied it to ecological analysis. Because of such borrowings, we begin our historical path-taking with a prefatory comment on the social sciences.

3

As the social sciences emerged as freestanding disciplines in the nineteenth century, they struggled with the problem of how much human behavior should be attributed to our biological nature versus social learning. Obviously, humanity shares characteristics with the animal kingdom, particularly the large nonhuman primates. At the same time, there is a sense of great difference. Depending on which social scientist one consults, causal priority seems to shift from "nature to nurture." According to some, genes, anatomy, or chemistry determine human behavior. For others, human behavior is determined by norms, moral values, or the mind and linguistic constructs, demography, or God's grand design. Human ecology is particularly at risk in such discussions, because it attempts to account for environmental variables and biological predispositions and to merge these with social variables unique to humans such as symbolic language, elaborative normative systems, values, and meanings.

The sociologist Pitirim Sorokin, in criticizing the application of biological analogies to human society, captured the reality of social science attempts to include the biological domain. He noted that "sociology has to be based on biology; that the principles of biology are to be taken into consideration in an interpretation of social phenomena; that human society is not entirely an artificial creation; and that it represents a kind of living unity different from a mere sum of the isolated individuals."[3]

He went on to critique organicist, biosocial, geographic determinist, and demographic approaches to explaining human behavior and the patterns and processes of human society. To Sorokin, all such explanations suffered from too much dependence on analogy and too strong a desire for single causes. Yet each of the mainstream theories critiqued by Sorokin (and each of those that have emerged since) has had to find some rationale for attributing, incorporating, excluding, or compartmentalizing the priorities of environment, biology, and human culture. Each theory must assume that the observed regularities in human social life have an explanation.

Mainstream social theories have tended to cluster around certain key biophysical and environmental determinants. For example, the structure of a society and its processes of stability and change have long been attributed to "carrying capacity" levels as population presses against resource constraints (W. G. Sumner, A. G. Keller, Émile Durkheim, and W. R. Catton). Some may consider ecological processes and environmental conditions as aspects of symbolic systems (Louis Wirth and Walter Firey). Others may view the variety of human organizational patterns and processes as shaped by environmental variation (O. D. Duncan and Philip Selznick). Societal patterns and processes are mediated by adaptive technologies, for which the cultural elements must accommodate to technological change (William F. Ogburn and W. F. Cottrell). The structure of political power may determine, and, in turn, be shaped by, characteristics of natural resources.[4]

Our point is twofold. The first is that environmental sociology and its related kin of natural resource sociology are neither recent products of sociologists nor distinct from mainstream social theory. The second is that traditional mainstream social theory must make, either explicitly or implicitly, an accommodation to the dilemma of reconciling social and biological facts. An ecology that includes humans is like other zoological studies in that it begins with the biological and environmental conditions of the observed species, rather than a determined assertion as to how little such factors matter in explaining the observed behavioral patterns. Indeed, our stated goal is movement toward a unified theory of ecology that ultimately can account for the ecologies of all lifeforms, including humans.[5] A useful starting point is the ecosystem concept.

4

The ecosystem was formally defined by Sir Arthur Tansley in 1935. Several histories of the ecosystem idea have been published, notably J. Hagen's *An Entangled Bank* and Golley's *History of the Ecosystem Concept in Ecology*. Hagen writes as a science historian; Golley was a student of E. P. Odum and participated in the development of the ecosystem concept. Both limit their discussion to the rise of biological ecology and an ecosystem concept that largely excludes humans.

Re-reading Tansley's original argument is revealing. Tansley embedded his new ecosystem idea in a *festschrift* paper in *Ecology* entitled "The Use and Abuse of Vegetational Concepts and Terms."[6] Tansley was responding to Clementsian (that is, successional and organismic) attempts by South African ecologist John Phillips to develop concepts for understanding larger and larger biological entities. Phillips's "biotic communities" and "the complex organism" deeply dissatisfied Tansley on philosophical and scientific grounds. (In the 1930s, the linkage between philosophical and scientific reasoning was rather more accepted than in our own reductionist age.)

Tansley begins his article with a respectful genuflection to Cowles, the brilliant American botanist of succession, calling Cowles "the great pioneer." He then turns to Phillips and communicates his intentions: "Bluntness makes for conciseness and has other advantages . . ." Tansley then delivers a fierce attack on the logic and evidence of Phillips's article. It is in response to Phillips' holistic concept of "the complex organism" that Tansley explains his systems thinking and proposes the *system* concept:

> I have already given my reasons for rejecting the terms "complex organism" and "biotic community." Clements' earlier term "biome" for the whole complex of organisms inhabiting a given region is unobjectionable, and for some purposes convenient. But the more fundamental conception is, as it seems to me, the whole *system* (in the sense of physics), including not only the organism-complex, but also the whole complex of physical factors forming what we call the environment of the biome—the habitat factors in the widest sense. Though the organisms may claim our primary interest, when we are trying to think fundamentally we cannot separate them from their special environment, with which they form one physical system.
>
> It is the systems so formed which, from the point of view of the ecologist, are the basic units of nature on the face of the earth.[7]

Tansley's leap was to consider the sum of biotic and abiotic *components* or parts (he used both terms interchangeably) as an integral unit, at the same time avoiding either false primacy of biota or the philosophical speculations of "organismic wholes."

Less revealed by the historians of ecology is how Tansley—in the same article—went further toward a *human* ecology. In the final section of the *festschrift* paper, he responded to Phillips's definition of biotic factors:

> It is obvious that modern civilized man upsets the "natural" ecosystems or "biotic communities" on a very large scale. But it would be difficult, not to say impossible, to draw a natural line between the activities of the human tribes which presumably fitted into and formed parts of "biotic communities" and the destructive human activities of the modern world. Is man part of "nature" or not? Can his existence be harmonized with the conception of the "complex organism"? Regarded as an exceptionally powerful biotic factor which increasingly upsets the equilibrium of preexisting ecosystems and eventually destroys them, at the same time forming new ones of very different nature, human activity finds its proper place in ecology.[8]

In one elegant stroke, Tansley integrates humans into ecology, identifies the potential failure of separating "natural" from "unnatural" biotic change, and presciently warns of the potential destructive impact of human activity. Tansley formally defines (with the same emphasis as he placed earlier on *ecosystems*) such human-dominated activities as *anthropogenic ecosystems*.

Tansley then summarizes his rebuttal of Phillips and Clementsian holism, codifying his concepts in an "essential framework." Even here, there is a revelatory comment on dynamic equilibrium. Tansley sums up his ecosystems idea: "The fundamental concept appropriate to the biome considered together with all the effective inorganic factors of its environment is the *ecosystem*, which is a particular category among the physical systems that make up the universe. In an ecosystem the organisms and the inorganic factors alike are *components* which are in relatively stable dynamic equilibrium. Succession and development are instances of the universal processes tending towards the creation of such equilibrated systems."[9]

Tansley thus deals with many of the contemporary issues surrounding the structure of human ecosystems: systems thinking versus vague organicism, the co-equality of biotic and abiotic factors, the potentially dominant and integral role of humans in the fate of the system, and the reality of dynamic equilibrium states that are common within such ecosystems.

Tansley's concept codified the key parts of ecosystems, giving name and shape to growing efforts to describe the structure of complex biological entities involving multiple species, flora and fauna, and abiotic conditions. A second phase in the development of the ecosystem concept focused on fluxes or flows *within* ecosystems. It was led by an extraordinary pairing of an elder mentor and intense young scientist.

5

In 1941, G. Evelyn Hutchinson (a limnologist) was joined at Yale University by a recently minted Ph.D. Raymond Lindeman. The story of Lindeman's brief career (he died at age 27) and of the initial rejection and ultimate acceptance of his now-classic work has been told by Golley, Hagen, and most intriguingly by Robert E. Cook.[10]

Lindeman's paper (the last chapter of his doctoral thesis) was published in *Ecology* under the title "The Trophic-Dynamic Aspect of Ecology." Lindeman grasped that Tansley's concept could enable ecology to integrate both biotic and abiotic factors *and* link structure and process in ways that explained dynamics of ecological succession and change. Lindeman's definition of ecosystem modernized Tansley's description and introduced *process* as fundamental: "The *ecosystem* may be formally defined as the system composed of physical-chemical-biological processes active within a space-time unit of any magnitude, i.e., the biotic community *plus* its abiotic environment. The concept of the ecosystem is believed by the writer to be of fundamental importance in interpreting the data of dynamic ecology."[11]

Focusing on food-cycle relationships, Lindeman laid out a largely theoretical and generalized framework. *Ecology*'s editors initially rejected the paper on the grounds that it was too theoretical and reconsidered only after intervention by Hutchinson. Figure 4.1 is from that paper and is remarkable for its integration of parts and flows, identification of trophic levels, and its central variable—"ooze." Lindeman, as

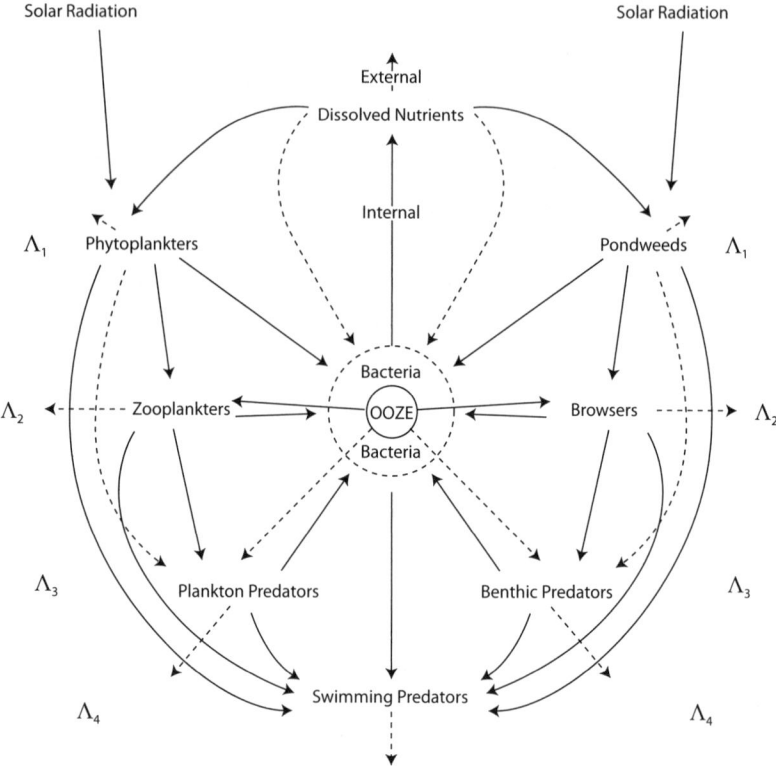

Figure 4.1. Generalized food-cycle relationships (from Lindeman, "The Trophic-Dynamic Aspect of Ecology," 1942)

a naturalist deeply familiar with his studied portion of nature, understood the "messiness" of real ponds and lakes and its implication for ecological theory: "The difficulty of drawing clear-cut lines between the living *community* and the non-living *environment* is illustrated by the difficulty of determining the status of a slowly dying pondweed covered with periphytes, some of which are also continually dying. As indicated in figure 4.1, much of the non-living nascent ooze is rapidly reincorporated through 'dissolved nutrients' back into the living 'biotic community.' This constant organic-inorganic cycle of nutritive substance is so completely integrated that to consider even such a unit as a lake primarily as a biotic community appears to force a 'biological' emphasis upon a more basic functional organization."[12]

Lindeman's work initiated a burst of ecosystem studies—primarily energy analyses and nutritional studies. A slow ascendancy of the ecosystem concept occurred, overshadowing rival terms and discarding the dead-end organicism reviled by Tansley. Interest was fueled by advances in general systems theory and the fruits of Lindeman's trophic-dynamic framework.[13]

A third phase in the development of the ecosystem concept was initiated with the publication of Eugene Odum's textbook *Fundamentals of Ecology*. Textbooks as a "vernacular expression" of science can be revealing and important, as Gould describes: "To learn the unvarnished commitments of an age, one must turn to the textbooks that provide 'straight stuff' for introductory students. Yes, textbooks truly oversimplify their subjects, but textbooks also present the central tenets of a field without subtlety or apology—and we can grasp thereby what each generation of neophytes first imbibes as the essence of a field."[14]

Odum was driven to distinguish ecology from its precedent sciences of botany and zoology. To do so, he needed an overarching framework and an identifiable unit of analysis. He introduces the ecosystem concept to students by first repeating Tansley's integrative strategy: "Living organisms and their nonliving (abiotic) environment are inseparably interrelated and interact upon each other. Any area of nature that includes living organisms and nonliving substances interacting to produce an exchange of materials between the living and nonliving parts is an ecological system or ecosystem."[15]

He then quickly builds a taxonomy: Ecosystems include *abiotic substances, producers, consumers,* and *decomposers*. He declares the ecosystem as "the basic functional unit in ecology" and then moves on to explain the breadth of its implications: "The concept of the ecosystem is and should be a broad one, its main function in ecological thought being to emphasize obligatory relationships, interdependence, and causal relationships. Ecosystems may be conceived and studied in various sizes. A pond, a lake, a tract of forest, or even a small aquarium could provide a convenient unit of study. As long as the major components are present and operate together to achieve some sort of functional stability, even if for only a short time, the entity may be considered an ecosystem."[16]

Odum's text, reprinted and revised several times, helped train a generation of ecologists. Ecosystem ecology expanded beyond easily bounded lakes and ponds and became increasingly quantitative, particularly in its energetics and kilocaloric measurements. The Hubbard Brook studies led by Likens and Bormann, so critical to the development of the HEM, examined cycling of nitrogen within an entire watershed.[17] The Hubbard Brook Research Program expanded empirical evidence in ecosystem functioning at the same time as training a new generation of ecosystem ecologists.

A critical element of these earlier ecosystem concepts was the treatment of structure as a set of functioning parts. Ramon Margalef, in his brief but influential paper "On Certain Unifying Principles in Ecology," suggests the broader conception of emergent structure employed in our model of human ecosystems:

> Ecosystems have a structure, in the sense that they are composed of different parts or elements, and these are arranged in a definite pattern ... the 'real' structure of an ecosystem is a property that remains out of reach, but this complete structure is reflected in many aspects of the ecosystem that can be subjected to observation: in the distribution of individuals into species, in the pattern of the food net, in the distribution of total assimilatory pigments in kinds of pigments, and so on.
>
> Structure, in general, becomes more complex, more rich, as time passes; structure is linked to history. For a quantitative measure of structure, it seems convenient to select a name that suggests this historical character, for instance, maturity. In general, we may speak of a more complex ecosystem as a more mature ecosystem. The term maturity suggests a trend, and moreover maintains a contact with the traditional dynamic approach in the study of natural communities, which has always been a source of inspiration.[18]

A fourth phase in the ecosystem concept's development followed from Margalef's insight and was led by Odum's brother H. T. Odum.

6

H. T. Odum had worked on portions of his brother's *Fundamentals of Ecology*. His own research focused on steadily more detailed energy analyses of complicated ecosystems. H. T. Odum's *Systems Ecology: An Introduction* codified ecosystem energetics, complete with its own notational system. He emphasized the open-systems nature of ecological entities. Borrowing his terminology from general systems theory and grounding his ideas in a natural historian's realism, Odum "engineered" the necessary scope of the ecosystem concept: "An organized system of land, water, mineral cycles, living organisms, and their programmatic behavioral control mechanisms is called an *ecosystem*."[19]

To H. T. Odum, an ecosystem necessarily included its "input environment," the delimited "system," and its "output environment." This allowed the ecologist to avoid false closure of ecosystem boundaries and to link ecosystems to each other and to larger biomes through measurement of input/output flows.

Odum's text, along with his brother's earlier work and that of their students, largely trained a generation of American biological ecologists and solidified the ecosystem concept within ecology. It also helped popularize the term *ecosystem* for a burgeoning environmental movement and popular media. Yet, over time, a disconnect between the machine systems analogy and the research of a new generation of ecosystem ecologists began to grow. The ecosystem concept was critically re-examined.

7

On August 9, 1999, Robert V. O'Neill presented the Robert H. MacArthur Award Lecture (named after the renowned Canadian-born American ecologist) at the Ecological Society of America's annual meeting in Spokane, Washington. The provocatively titled lecture "Is It Time to Bury the Ecosystem Concept?" was an effort to reconcile the "machine analogies" of the traditional ecosystem concept with the emerging and much more chaotic knowledge of how ecosystems actually function.

O'Neill began his critique with a reminder of how the idea of "ecosystem" is a scientific construction: "The simple fact is that the ecosystem is not *a posteriori*, empirical observation about nature. The ecosystem concept is a paradigm [*sensu* T. S. Kuhn], an *a priori* intellectual structure, a specific way of looking at nature."[20]

O'Neill then described the "vigorous backlash" to use of the ecosystem concept, resulting from the "apocalyptic focus" of the environmental movement. He memorably noted that "ecology oversold its ability to predict doom" and criticized the mythic nature of an integrated, homeostatic ecosystem in equilibrium. He then turned to a discussion of scientific problems, beginning with definitions: "At present, the terms *ecosystem* and *ecosystem theory* are used in many different ways. At one extreme, *ecosystem* is a convenient term, relatively free of any assumptions, that indicates the interacting organisms and abiotic factors in an area. At the other extreme, *ecosystem* is a precisely defined object of a predictive model or theory."[21]

Ecosystems have spatial boundaries (for example, a watershed, pond, or marshland) within which most ecosystem structure is observed. Yet the "component populations" of an ecosystem may have much broader distributions. The ecosystem concept assumes some level of spatial homogeneity, yet internal *heterogeneity* is an essential feature of any ecosystem's structure and is crucial to predictive understanding. Ecosystem analysis often begins with a comprehensive species list, yet the level of species substitutability that can be accepted without declaring the ecosystem a completely different system is frustratingly unclear. O'Neill approaches the problem logically: "The inconsistencies are brought out by the seemingly inane question: 'Do ecosystems die?' Consider, for example, a northern lake that has undergone eutrophication. If the ecosystem is defined as a functional system at a spatial location, then the lake is the same ecosystem, albeit altered by changed conditions. On the other hand, if the ecosystem is defined by the species list, then the oligotrophic ecosystem has been killed and replaced by eutrophic ecosystem."[22]

O'Neill then critiques the problems of stability, its definition (particularly in the face of species substitutability), and the need to treat "disturbance regimes" such as wildfire or extreme drought as part of ecosystem structure.

The topic of disturbance regimes leads O'Neill to a discussion of *Homo sapiens* and its species role in ecosystems. His scope is limited to biological ecology and his language is colorful: "The ecosystem concept typically considers human activities as external disturbances to the ecosystem. Other invasive pests, such as kudzu and brown rats, are considered as ecosystem components, and their impact on structure and function considered explicitly. *Homo sapiens* is the only important species that is considered external from its ecosystems, deriving goods and services rather than participating in ecosystem dynamics. If there was ever a species that qualified as an invasive pest, it is *Homo sapiens*."[23]

O'Neill then begins his proposal for assembling a "new" ecosystems paradigm, which he suggests is merely "putting splints and patches on an old horse." He recommends that ecosystems be defined as including multiple spatial scales, from the localized system to the potential dispersal range of all species within the local system. Because human disturbance often affects dispersal range, *Homo sapiens* becomes a keystone species that alters the structure of ecological systems (as O'Neill notes, "like the beaver"). He describes several issues related to stability and then asks: "Is it time to bury the ecosystem concept? Probably not. But there is certainly need for improvements before ecology loses any more credibility. . . . Perhaps the most important implication involves our view of human society. *Homo sapiens* are not an external disturbance, they are a keystone species within the system. . . . Certainly, we don't want to dismiss the current theory prematurely. But we must understand that the machine analogy is critically limited."[24]

In a concluding statement with its equestrian metaphor intact, he calls for an alternative model of ecosystems that reflects both biological complexities and the role of humans: "It would not be wise to send the old dobbin to the glue factory before we determine how well the new one takes the bit. But it certainly seems to be time to start shopping for a new colt."[25] Biological ecologists from Tansley onward have thus elaborated the ecosystem concept, touching on the role of humans but largely avoiding the task of describing a *human* ecosystem. A parallel genealogy exists in the social sciences.

F • I • V • E

The Roots of Human Ecology

1

The roots of a human ecology lie primarily in general ecology, sociology, geography, and anthropology, as documented by numerous literature reviews. The idea for the application of general ecological principles to human activity was sparked by sociologists at the University of Chicago in the 1920s and 1930s. Sociologists Robert E. Park and Ernest W. Burgess and others drew analogies between human and nonhuman communities, describing society's symbiotic and competitive relationships as an organic web. Burgess proclaimed that "the processes of competition, invasion, succession, and segregation described in elaborate detail for plant and animal communities seem to be strikingly similar to the operation of these same processes in the human community." Lewis Mumford, in his magisterial *Technics and Civilization*, elaborated on the role of cultural choices and technology in such systems.[1]

Biological concepts such as "competition," "commensalism," "succession," and "equilibrium" were freely borrowed, mirroring the biologists' use of social science concepts. Usage often drifted between analogy and metaphor.[2] Borrowing from contemporary plant ecologists and their focus on "plant community zones," these early human ecologists moved from classrooms to city streets to map "natural areas" or zones of the urban metropolis.

The Chicago School produced a series of impressive studies that combined ethnographic, humanistic, and statistical techniques. They recorded the psychological, social, and geographic dimensions of the giant city of Chicago arising in the American heartland. William I. Thomas and Florian Znaniecki's *The Polish Peasant in Europe and America*, Harvey Zorbaugh's *The Gold Coast and the Slum*, Frederick Milton Thrasher's *The Gang*, Clifford Shaw et al.'s *Delinquency Areas*, R. Faris and H. Dunham's *Mental Disorders in Urban Areas*, and Wirth's classic *The Ghetto* were major studies.[3]

As the analogies were pressed further, the methodological difficulties of applying these biological concepts quickly emerged. W. W. Weaver's monograph *Philadelphia: A Study of Natural Social Areas* challenged the usefulness of the "natural area" concept; F. A. Ross's paper "Ecology and the Statistical Method" provided a penetrating, if brief, analysis.[4]

It fell to a student of R. M. MacIver's, Milla Aïssa Alihan, to critique the ecological school of American sociology comprehensively. Her too-little-read book *Social Ecology: A Critical Analysis* is a patient, skilled, and fearless analysis; her work stands as a model of logical inquiry and thoughtful evaluation of social theory.[5] (Not all would agree; O. D. Duncan characterized the book as "a polemic.") Alihan began *Social Ecology* by describing "the doctrine and its setting," describing human ecology as practiced in the 1930s as "essentially American." She saw American sociology reacting to the perceived need to develop independence from European thinkers.

Alihan described the Chicago School's publications (several of which are listed above) as "factual studies" and examined their relationship to the theoretical concepts used by the early human ecologists. She found the concepts vague and the analogies distorted. Her criterion was scientific usefulness: "in human ecology the concepts 'succession,' 'dominance,' and 'natural area' are so all-inclusive and general that their application results in distortion as well as confusion. Surely no scientific method requires such straining interpretation in the interests of simplification."[6]

Focusing her critique first on the ecological version of "community" (the concept of *ecosystem* had not yet been formulated by Tansley), Alihan moved on to analyze other "specific borrowings." In each case, the borrowed biological concepts were expanded to include sociological concerns, with dizzying results. Clements's concept of plant succession is an example: "'Succession' . . . has been applied to the displacement of racial, age, economic, or cultural groups, institutions, utilities, structures, cultural factors, architectural styles; to sequences of technological inventions and cultural trends; and in short to anything and everything."[7]

To Alihan, the human ecological approach taken by the Chicago School had reached an intellectual dead end, and the lack of unity between theory and facts, doctrine and methods, and approach and conclusions had grown all too apparent.

2

The limited focus on spatial relationships and urban life and the over-extension of biological metaphors eventually led to a search for a more suitable framework for human ecology. That search, active in the 1950s and 1960s, led to the POET model (see figure 5.1). This model defined the "ecological complex" as the interaction among population, organization, and technology in response to the environment.[8] These were to be human ecology's "master variables"; their interaction was to be the human ecologist's central concern.

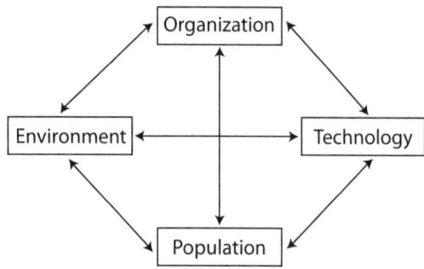

Figure 5.1. The POET model

A derivative "IPAT" model modified the interactions to estimate environmental impacts as a function of populations, affluence, and technology. Other acronym-favoring efforts were made to identify key variables; an example is Kenneth D. Bailey's addition of "level of living" and "information" to the model, creating PISTOL as a more elaborate "ecological complex."[9]

The description of the "human ecosystem" as a working construct proceeded in halting steps, mirroring the emergent use of Tansley's concept in biological ecology. In 1955 (about the same time as Odum's ecology text popularized "ecosystems" to a generation of American graduate students in biology), L. R. Dice wrote *Man's Nature and Nature's Man: The Ecology of Human Communities*. Dice attempted a wary synthesis: "The major thesis of this book is that man is able to exist only as a member of some community in which associated species of plants and animals supply him with food and other essential supplies and services. Man, in turn, has profound effects upon his plant and animal associates. These interrelations between man and other kinds of organisms are of great significance for human welfare."[10]

Cautiously, Dice started with a kind of conceptual glossary, describing the concepts of ecosystem, human community, society, culture, and adaptation. He then began to assemble the relationships between ecological and social factors. Echoing Tansley's use of North American grazing, Dice provides a practical resource management example:

> In the mountains of eastern Oregon, domestic livestock come into competition for summer forage with the introduced American elk. The range will support only a certain amount of pasturing. If elk increase, then fewer domestic cattle, sheep, goats, and horses can be supported. In this particular situation, many diverse economic and sociologic relations are involved. The interests of foresters, livestock raisers, farmers, sportsmen, and nature lovers are all affected. Even the local merchants, manufacturers, truckers, teachers, preachers, lawyers, and other city dwellers are concerned. The interests of the whole county, and to a lesser extent those of the state

and nation must also be considered. The adjustment of these complex and sometimes conflicting interests is the province of the political administrator. In the past, the reconciliation of such conflicts has all too often been attempted without an adequate consideration of the ecology of the area. Administrators, however, are becoming increasingly aware of the importance of scientific advice from range managers, soil conservationists, foresters, wildlife managers, and other kinds of practical ecologists.[11]

By chapter 15 of his book, Dice was ready to synthesize. The chapter, titled "Human Ecosystems," begins with the simple act of broadening the "human community" concept: "A human community consists not only of men, women, and children, but as a minimum, it includes also certain plants that supply food to man, either directly, or indirectly through animals. Every human community is actually far more complex than this; it includes food chains of numerous kinds, domestic animals and perhaps crop plants, parasites, diseases, and numerous other species of organisms, some of which serve as regulatory mechanisms or perform other ecologic functions."[12]

Dice followed this track, describing (in the language of the time) that where there is a "sparse human population in a low state of culture," man is simply one species with a natural community, and part of a "natural ecosystem." Close to synthesis, he both introduces the concept and diverts to typology: "A more comprehensive classification of human ecosystem is greatly needed."[13]

While the subsequent classification system is intriguing (including such gems as "camp-centered tribal ecosystems," "homestead ecosystems," and more), the synthesis of ecological and sociocultural variables that was so close to achievement was, unfortunately, not consummated.

3

Human ecology was not limited to the Chicago School or American sociology or for that matter sociology. Other disciplines and intellectual interests were at work on the ecology of humans. Marston Bates provided a comprehensive summary, documenting at least five divergent ways in which ecology was being applied by social scientists: a human ecology stemming from medicine, stressing the relationship of environment and disease; a human ecology derived from geography; the American Sociology School; a broad "tag line" to indicate studies of universal concerns; and human ecology stemming from anthropology. Cultural anthropologists moved toward modern ecology, with energy flow being the synthesizing link. Roy Rappaport's *Pigs for the Ancestors* captured both the ecological and social complexity of the Tsembaga's use of resources as well as the clarifying importance of energy and nutrient flow in human systems.[14]

Geographers contributed as well, often providing an urban and planning perspective across a range of spatial scales. James O. Berger carefully examined the planning process using the Pinelands National Reserve in southern New Jersey as a case study, illuminating the limitation of ecological planning devoid of sociocultural knowledge. Frederick Steiner described planning and design methods that synthesized social and natural sciences.[15]

Yehudi Cohen's three-volume set of readings brought together much of the literature on biosocial analysis and emphasized the importance of adaptive strategies. A substantive advance in anthropology's contribution—which had a powerful effect on the present authors—was J. W. Bennett's *The Ecological Transition: Cultural Anthropology and Human Adaptation*. The book was largely dismissed by mainstream anthropology as marginal when it first appeared. Bennett presented a paradigm of human ecology or cultural ecology—he explicitly treated these subfields as "virtually identical." Figure 5.2 was and still is revelatory. It mapped a synthesis of social and biological phenomena both realistic and consequential. His concept of "Presses" described values, desires, needs, and wants; their spiraling and deeply psychological demands for consumption were a (not *the*) driving force toward resource degradation. Bennett understood that "Presses" can dampen wants and needs as well (ritual taboo can serve this function, for example) but saw this as a specialized characteristic at best: "If such subliminal regulation exits in the human ecological system, it has not been historically

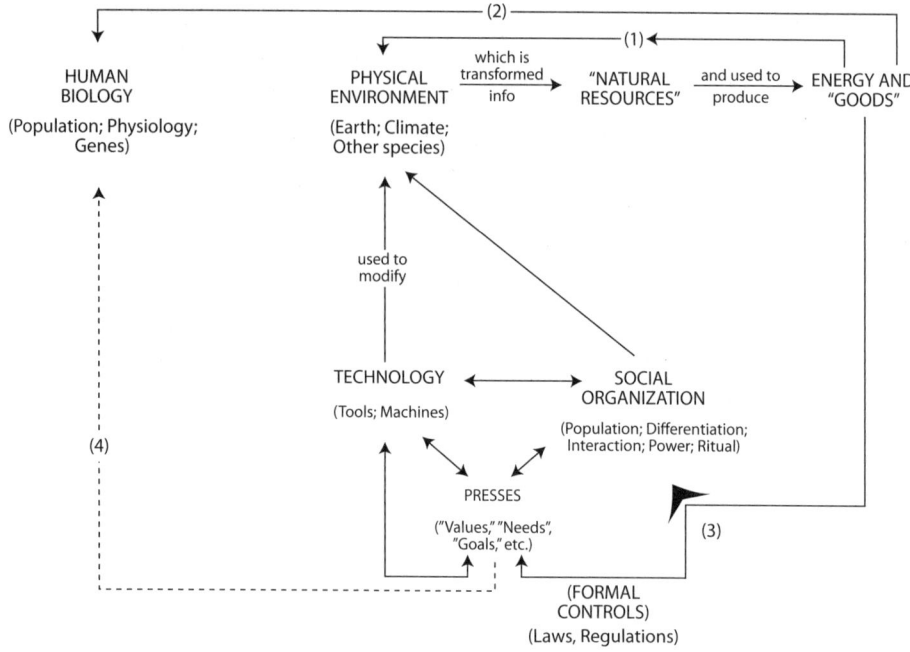

Figure 5.2. A paradigm of human (or cultural) ecology, emphasizing the output function (from Bennett, *The Ecological Transition*, 1976)

sufficient to preserve the system from abuse and overuse of resources. That is, there is no evidence of automatic controls operating in the system consistently in all societies."[16]

Bennett's bias was toward the cultural—how the strategic actions of people and organizations within specific cultures influence nature. Yet if culture was the master variable, Bennett's insistence on a policy-relevant human ecology repeatedly brought his argument toward synthesis of biological and sociocultural systems. Prescient but largely ignored, *The Ecological Transition* deeply influenced our present effort.

Anthropologists continued to work toward a culturally bounded ecology and the use of ecosystem concepts (see for example the work of Roy Ellen and E. F. Moran).[17] Some focused on a dominating cultural frame of reference, where "resources" were culturally defined. Others were biological, where human physiological adaptation (such as high-altitude lung capacity) influenced behavior. Geographers contributed, from W. G. East's intriguing and only superficially deterministic *The Geography behind History* to the massive and influential collection *The Earth as Transformed by Human Action*.[18]

As the various social sciences advanced their proposed partnership with the biological sciences, numerous pathways were explored—sociobiology, natural resources sociology, urban ecology, and environmental economics are examples. Most recently there has been a return to the systems thinking of the 1970s couched in the language of "coupled natural-human systems." The ecosystem concept was an occasionally attractive tool, and multiple efforts were initiated. Some of these projects came from theory, such as H. E. Kuchka's *The Method for Theory: A Prelude to Human Ecosystems*. Others came from attempts to apply the ecosystem idea to manage resources, economic development programs, sustainability initiatives, and other worthy goals.[19]

In 1997, we published a brief paper "The Human Ecosystem Part I: The Human Ecosystem as an Organizing Concept in Ecosystem Management." The second part of the paper was published simultaneously.[20] Borrowing a portion of the title from Tansley's original work, the paper outlined—in explicit terms—what we considered the essential features or key variables of human ecosystems.

4

A small, insightful literature emerged around the HEM. Thomas Rudel offered an early critique, calling for a more anthropocentric approach: "The chief problem with the framework [model] is that it does not put people as knowledgeable agents, at the center of the human ecosystems."[21]

Valerie Luzadis and her colleagues provided a more thorough assessment, urging that the model be revised to better represent processes and flows, particularly flows of energy and information; to more explicitly integrate social and biological patterns and processes; and to further emphasize the biophysical aspects of the system. We hope this present volume addresses these issues. Graduate students (particularly at Yale

University and the University of Idaho) have continued to test, apply, and critique the model, providing both provocative questions and worthy recommendations.[22]

The model has been applied by others (sometimes only peripherally) to a variety of problems: policy-making in Vermont, greenways in Texas, resource management in Australia. It has also been used to examine health issues and forest-dependent communities in Canada. In preparation for the World Summit on Sustainable Development in 2002, the International Council for Science proposed the model as an alternative framework for "state of the environment" reporting.[23]

An example of its general application is the Sonoran Institute's *Social Indicators for Sonoran Desert Ecosystem Monitoring*.[24] Covering 55 million acres, the Sonoran Desert bioregion includes parts of the U.S. states of California and Arizona and portions of the Mexican states of Baja California and Sonora. The region is noted for its importance to pollinators, migrating bird species, and high biodiversity, particularly endemic plants, reptiles, and fish. The bioregion includes a burgeoning human population and the metropolitan centers of Phoenix and Tucson, Riverside County in California, and the Mexican cities of Hermosillo and Mexacali. With approximately half of the region in the United States and half in Mexico, bioregional management requires both an ecological and cross-cultural strategy. The Sonoran Institute's report applied the model directly to organize the work they did to foster adaptive management. The HEM was considered in the design of the Sonoran Desert ecosystem monitoring framework and in evaluating proposed social indicators. The report goes on to describe a workshop for selecting specific social indicators and provides examples in map, table, and graphic form. The model—applied as an "organizing concept" and framework for data collection—is echoed throughout the report.

We have also continued our experiments and applications. Burch, working closely with Morgan Grove (a former graduate student of Burch's) and Stewart Pickett of the Institute of Ecosystem Studies, has applied the HEM to help organize their long-term study of urban ecosystems. Baltimore, Maryland, is a key study site. The model is considerably modified, with an expanded set of biophysical variables and a list of ecosystem processes and patterns. A recent volume (with Burch and

Machlis as coauthors), *The Baltimore School of Urban Ecology: Space, Scale, and Time for the Study of Cities,* summarizes these efforts.[25]

Through United Nations University in Tokyo and several scientific/municipal organizations in China, Machlis, Burch, and others have introduced the model to urban managers from throughout Asia: from Kuala Lumpur to Beijing to Ho Chi Minh City, and the model has been used as a tool for organizing single-resource accounting (in this case, the waste system) for Bangkok and Dhaka.[26]

The model has had other applications as a tool for quickly conducting interdisciplinary assessments and developing mid-term scenarios during major environmental disasters. Machlis used the HEM as an organizing framework and operational model during the 2010 Deepwater Horizon oil spill and Hurricane Sandy in 2012. In both cases, the model's integrated use of biophysical and sociocultural variables allowed scientists from disparate disciplines to share insights rapidly and predict "cascading consequences" of these disasters.[27] A detailed description is in chapter 8.

Adjustments to the model—in response to both the queries of managers and students as well as accumulating knowledge—have continued, with the most recent version in place for this book. Hence, the HEM that we have attempted to construct is both the product of long-term efforts by many and a continually evolving project.

There have been only seventy years, the comfortable span of a modern lifetime, between Tansley's *festschrift* definition of an "ecosystem" and our current exploration. The intellectual genealogy just outlined is largely our own, far from comprehensive. We have described what intrigued us and our path of discovery. We now turn to describing in more detail the HEM and, in doing so, the structure and dynamics of human ecosystems.

S · I · X

Key Components and Variables for Analyzing Human Ecosystems

1

Chapter 2 outlined the HEM, and figure 6.1 lists the basic key components and variables for describing, analyzing, and working with human ecosystems. This chapter draws on lessons learned from prior application of the model and includes those lessons in a modest revision of the early work. We detail each component and variable by providing a general definition and description, suggesting ways that the variable can be measured, and giving selected examples of how a variable may influence other variables of the human ecosystem. The passage of time and the application of the model in many situations have left most of those early components and variables intact, with only a few key additions and deletions. Our goal remains to be as inclusive as necessary and as parsimonious as possible but still get the job done. We seek to provide the HEM user with an integrated set of social, economic, and ecological measures that can be collected over time. Most can be derived from available data sources, are grounded in theory, and are useful for ecosystem management and decision-making.

The stability comes from the components remaining constant, regardless of the problem addressed. The variation comes with the ability to use some variables and not others and the ability to add variables that

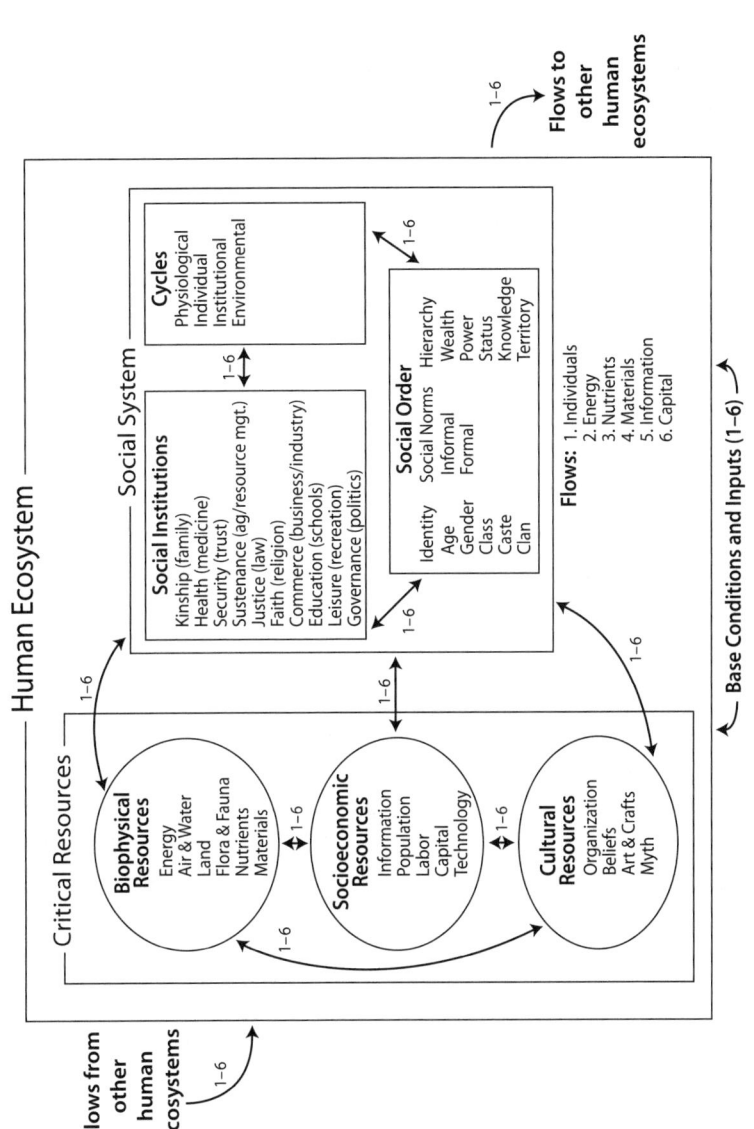

Figure 6.1. The Human Ecosystem Model

seem necessary for the utility of the model. Components cluster those variables that together define a similar conceptual unit or focus. As a New England county is made up of several townships and as Louisiana is a collection of parishes, our HEM components combine a cluster of variables that share a similar focus. The "biophysical resources" grouping differs in a consistent way from the "cultural resources" component. The utility for the analyst is one of scale. The inclusive unit can be compared to another inclusive unit or specific variables within the components can be compared.

2

The *biophysical resources* component combines those basic drivers shaping the nature, types, and persistence of lifeforms and ecosystems. This component is composed of eight critical variables.

Energy is the ability to do work and/or create heat. Energy is a critical natural resource, and its influence on social systems is well documented. As Cottrell noted, the energy available to humans "limits what we can do, and influences what we will do."[1] Energy flows vary by type of source (hydroelectricity, petrol, natural gas, solar, nuclear, wood, and so forth) as well as quality (high- or low-entropy) and flow (continuous, cyclical, or interruptible). An important element is the locus and scale of control (external or internal, local or global, multinational or household). Energy can be measured by heat output (kcal) or economic value ($/kcal). Changes in energy flows can dramatically alter social cycles and the social order (witness the North American oil shortages in 1973 and 1979 and the overstocks and low prices of 2014–15) and can force social institutions (such as the recreation industry or agriculture) to make significant adaptations.

Land includes terrestrial surface, subsoil, and underground features and is a critical resource for both its economic and cultural value. It can be characterized by ownership patterns (public or private), cover (vegetation or plant community types), use (agricultural, forestry, urban, and so forth), and economic value. Changes in land use can often be measured in hectares by land cover-land use type. Such changes often follow restricted and predictable trajectories, as forested land is altered to agricultural and then urban uses. Other trajectories, such as the urban abandonment of Detroit, are similarly predictable.[2]

Land use is regulated by systems of rights and responsibilities assigned to certain parts of the land. In some settings different elements of land have different ownership patterns. In the New England states, one owner may own the land or shore of a water body, another may own the land under the water, and another owns the water supply. In the arid western states ownership of water and the surrounding land commonly follows the principle of first in time, first in right. In the Midwest a farmer may own the rights to the surface of a landscape, but some other entity owns the subsurface rights. Clearly the norms regarding rights and responsibility for land have great influence on many social institutions (sustenance and commerce are examples), and changes in land use often are reflected in altered hierarchies of wealth, power, and/ or territory through shifts in land tenure and property rights.

Water includes surface, subsurface, and marine supplies. Ground water (quickly renewed) and aquifers (a form of capital stock not easily renewable) can both be integrated into human ecosystems. Water resources can be characterized by quality, flow (acre-feet/second), distribution patterns, and cyclical trends (wet years or drought periods). The control and distribution of water constitute a major source of economic, social, and political power.[3] Changes in water quality can impact social institutions such as health and commerce; water rights are crucial to maintaining social order; access to water influences wealth.

Water resources interact with land and influence landscape. Western China in Gansu Province and the central part of the North Island in New Zealand along with other regions in the world have large loess soil deposits in combination with high fluctuations of wet-dry cycles that restrict certain agricultural practices. The Yellow River in China gains its name from the movement of these grainy sediments in the river. In New Zealand the high unemployment of the Great Depression combined with the lack of certain nutrients in its loess soils led to the creation of one of the largest human-made forests, where radiata pine imported from central California transformed this landscape into a major economic source in the late 1950s.

Air has twenty or more definitions in the dictionary. Some, like "air your complaints," are clearly not part of our lexicon. Air best serves our purposes in combination with water. Airsheds influence water, such as acid rain where coal fired energy plants in the U.S. Midwest created rain that was a factor in killing fish and other life in the lakes of the Northeast.[4] Similar airsheds mixed acid from the power plants and rain in middle Europe, which flowed along the airshed and dropped the solution on lakes in Scandinavia, killing freshwater shrimp and disrupting traditional cultural events around the harvest of the shrimp.

Airsheds and watersheds are necessary analytic categories as they provide a biophysically determined set of boundaries that clearly interact with variables in socioeconomic and cultural resource systems. Certainly the distribution of radiation from nuclear power accidents or testing ranges is transmitted by both the air and the watersheds of great rivers like the Columbia in North America and via underground water networks, contaminating drinking water.

An important measure of variation in the influence of land, air, and water is the albedo, which measures the total radiation from the sun that is reflected by a surface. This measure that is now monitored by satellite imagery can show differences in albedo of water, sand, grassland, forest, new snow, and so on. Changes in the albedo of different sources can be compared to one another or to the whole earth or to other planets. For instance, the conversion of a large section of the Amazon rainforest to pasture will be recorded in the changed albedo of the earth. The interplay between human actions and biophysical responses can be measured in albedo values, a long term and significant measure.

Materials include basic products derived largely from natural resources. Examples include fertilizers (petrol as a source), dimension lumber (wood), silver and other minerals (ore), plastic (oil), glass, concrete, cocaine, and denim. The variety of materials used by human ecosystems varies by culture, stage of economic development, and consumption patterns. Common measures include economic value/unit and/or the flow of raw product (by ton, pound, ounce, or milligram). Much of the sustenance and commerce institutions are based on the production, distribution, and exchange of materials. When flows are altered, norms for use can be impacted (conservation incentives increase with price), and certain materials may be critical for specific institutions, such as precious gems for industrial use, or coca paste for the illegal drug trade.[5]

Nutrients include the full range of food sources used by a human population. The range of tolerance for nutrient gain or loss is relatively small in *Homo sapiens*, making food a critical resource on a continuous basis.[6] Food varies by culture (religious proscriptions may make certain foodstuffs inedible) as well as climate, and both the caloric value and nutritional supplies (such as amino acids) are critical. Modern human ecosystems include a wide range of imported foods (witness espresso coffee beans from Brazil being brewed in Montana gas stations), and few are self-reliant even for short, seasonal periods. The need for food resources certainly influences sustenance institutions such as agriculture. Food carries mythic connotations (the spiritual value of salmon to several indigenous tribes in the northwest; the turkey as a celebratory poultry). Hence, changes in nutrient flows can alter human health, social norms, and cultural beliefs.

Flora and *fauna* are critical resources beyond their function as nutrient and material sources; a wide range of flora have ecological, sociocultural, and economic value. Plants are vital sources of pharmacopeia, myth (the cedars of Lebanon and the redwoods of California are examples), and status (the American lawn).[7] Fauna, including domesticated livestock, pets, feral animals, and wildlife, have significant economic value through activities as wide-ranging as hunting, birdwatching, pet keeping, and (in some cultures) the production of aphrodisiacs. Flora and fauna can be valued biologically (such as species richness, number of endemic species, population size, genetic diversity), economically (dollar value per bushel, board foot, pelt, head, horn, or hoof), or culturally (proportion of citizens interested in preserving a species). Changes in flora and fauna, such as the threat of extinction or overpopulation, can lead to changes in nutrient supplies, myth, law, sustenance (particularly wildlife management and farming practices), and social norms toward the natural world.

3

The *socioeconomic resources* component clusters together those variables most critical for the emergence, adaptation, stability, and change of human ecosystems. We need to identify and describe the characteristics of the social mechanisms that interact with the biophysical and cultural resources and constraints to create and sustain the identity and processes of specific human ecosystems.

Information is a necessary supply for any biophysical or social system. Information flow (especially its potential for feedback) is central to general systems theory, sociobiology, and human ecology.[8] Information may be coded and transmitted in numerous ways: genes, "body language," oral traditions, electronic (digital data), print (local weeklies, national dailies, news magazines), film, radio, and television. It can be measured by both transmission rates (such as amount of local radio programming) and consumption patterns (newspaper circulation rates or number of subscribers to internet services). Information flow can significantly alter components of social systems, such as educational institutions or hierarchies of knowledge. The impact of information on other critical resources is also substantial (for example, the importance of maps in land management).

Population includes both the number of individuals and the number of social groups and cohorts within a social system. Population as a socioeconomic resource includes the consumption impacts of people, as well as their creative actions (accreting knowledge, engaging in sexual behavior, providing labor, and so forth). Human population growth is a dominant factor influencing much of human ecology and social systems, both historically and within contemporary nation-states, regions, and cities. Growth can be measured by natural increases (births over deaths/year) as well as migration flows. While population can act as an ecosystem stressor, it is also a supply source for many critical components within human ecosystems, such as labor, information (including genetic code), and social institutions.[9]

Labor has many definitions; in the HEM, it is defined as the individual's capacity for work (economists sometimes call this "labor power").[10] Applied to raw materials and machinery, labor can create commodities and is a critical socioeconomic resource. There are many measures: labor time needed to create a unit of economic value (hours/$100 value), labor value (measured in real wages), labor output (units of production per worker or hour labor), or surplus labor capacity (unemployment rates) are examples. Labor is critical to human ecosystems both for its energy and information content; that is, both relatively unskilled yet physically demanding labor (such as harvesting crops) and specialized, sedentary skills (such as resource planning or stock brokering) have economic and sociocultural importance. Changes in labor (such as increased unemployment) can impact a variety of social institutions and hierarchies from health care to income distribution.

Capital can have a range of meanings. A narrow definition treats capital as the "durable physical goods produced in the economic system to be used for the production of other goods and services." Other definitions include "human capital," "social capital," financial capital, and so forth.[11] In the HEM, capital is defined as the economic instruments of production, that is, financial resources (money or credit supply), technological tools (machinery), and resource values (underground oil). These instruments of production provide the basic materials for producing (with labor inputs) commodities. Capital is a critical socioeconomic resource; its influence over production, consumption, transformation of natural resources, and creation of by-products (pollution) is significant. Capital is often measured in dollar values, either for commodities produced or the stock of capital on hand. Changes in capital, either in its mix of sources (a new processing plant or mill organization) or output (a reduction in profits earned by the plant or mill or loss of resilience due to weakened organizations), can alter social institutions as well as hierarchies of wealth, class identities, and other features of the human social system.

Technology is the application of the methods and materials of applied science to achieve practical outcomes. It is both an opportunity and constraint that makes possible characteristic livelihoods, lifestyles, and legacies of specific human ecosystems. There is general agreement that specific technologies have shaped the characteristics of human ecosystems since the emergence of humans as a species twelve million years ago.[12] Humans lived as hunter-gatherers with primitive weapons and clever skills in organizing hunts and identifying wild foods. It was not until twelve thousand years ago that horticultural and pastoral societies emerged. Starting about five thousand years ago agrarian technologies permitted the creation of cities with a small proportion of the global human population. From 1750 to the present industrial technologies have been dominant, and only in the past few decades have some societies had the information and technologies to allow the emergence of consumer-based postindustrial societies. There may be considerable debate on the relative importance of technologies in shaping human ecosystems, but there is little doubt that technology has played a significant role in the lives of human ecosystems. Since the beginning of the twenty-first century social media technologies are having a dramatic, and as yet little understood effect on many components and variables of the human ecosystem.

4

The *cultural resources* component clusters together variables that determine values and perceptions about the human and nonhuman elements of human ecosystems. Our likes and dislikes, wants and desires, fears and strengths, action and inaction—all are shaped by the crucial variables in the cultural resources component.

In the HEM, *organization* is treated as a cultural resource, for it provides the structural flexibility needed to create and sustain human social systems. That is, our species' special ability to create numerous and complex organizational forms is a necessary skill in interacting with nature and society.[13] It is a *cultural* resource because there is demonstrated wide variation among cultures in how these generic organizing skills are employed. For example, citizens of the United States are willing to create, continually and often, new organizations to deal with collective issues: building a water supply system (irrigation districts), managing education (school boards), caring for the poor (welfare societies), and so forth. Organization can be measured by its diversity (the range of organizational types), intensity (the number of organizations), or saturation (the percentage of the population that claims membership). Organization is critical to natural resource management; ecosystem management is itself an experiment in new ways of organizing the relations between human and nonhuman domains.

Beliefs are statements about reality that are accepted by an individual as true; citizens may have the belief that forests are being overcut, that water quality is low, that certain salmon stocks may not be endangered. Beliefs differ from values, which are opinions about the desirability of a condition. Beliefs arise from many sources: personal observation, mass media, tradition, ideologies, testimony of others, faith, logic, and science. Beliefs are crucial to human ecosystem functioning, for they supply a set of "social facts" that individuals, social groups, and organizations use in interacting with the world. Hence, both environmental interest groups and industry associations rely on a public set of beliefs concerning environmental crises (which may or may not be factual) to energize and increase their membership and/or power. Beliefs can be measured by their ideological content (liberal or conservative), their intensity (the proportion of a population that feels strongly about a belief), and their public acceptance (the proportion of a population that shares a similar belief). As beliefs change, social institutions are often forced to respond. For example, changing public beliefs concerning the safety of nuclear power has led to a decline in nuclear power production in the United States.[14]

Art and crafts have been a part of human ecosystems since primates left the forest and moved to the savannah and created functional tribal groups. Stories and tales and guidelines for the present and future were carved on stones and painted on the walls of caves. These creative skills both represented the world as seen by the artist and contained lessons and legacies for present and future members of the community. In the modern era, the arts of sculpture, painting, photography, film, poetry, fiction, music, drama, and dance all continue the narrative of beauty and ugliness of delight and despair to be found in our lives and the world around us. Crafts apply the artistic impulse to the aesthetic of our daily lives—the design of chopsticks, the elegance of a cup or plate, the insight of a handmade wooden stool, the intuitive ease of a website. The crafts impulse is found in the effort and attention to making function attractive. It is a guide to the quality of life available in particular human ecosystems.

To the human ecologist, *myths* are narrative accounts of the sacred in a society; they legitimate social arrangements and explain collective experiences.[15] Hence, myths are an important supply variable because they provide reasons and purposes for human action. Myths are critical to human ecosystems as guides to appropriate and predictable behavior (witness Smokey Bear's admonitions about forest fire); they give meaning to and rationale for a wide range of social institutions and social ordering mechanisms. For example, the myth of "manifest destiny" provided U.S. citizens with a rationale for the permanent and private development of the American west; indigenous tribal groups simultaneously called on traditional myths to legitimate their role as temporary stewards of communal land. Myths operate at various scales: national myths (such as manifest destiny), community myths (a timber town's story of how and why it was founded), and clan myths (a family's story of its early matriarchs). Myths are difficult but not impossible to measure: festivals, symbols, legends are all indicators of myth supply. A change in myth (such as reduced perception of community self-reliance) can impact social institutions (such as faith) and a variety of social norms as well as resource use (such as wilderness).

Karen Armstrong's *A Short History of Myth* gives a clear understanding of the importance of myth throughout human history. It was myth that inspired the science that permitted humans "to travel through outer space and to walk on the moon, feats that were once only possible in the realm of myth." She uses the Neanderthal graves to "tell us five important things about myth. First, it is nearly always rooted in the experience of death and the fear of extinction . . . mythology is usually inseparable from ritual." She goes on to say that myths "force us to go beyond our experience . . . myth is not a story told for its own sake. It shows us how we should behave." Armstrong concludes that "all mythology speaks of another plane that exists alongside our own world, and that in some sense supports it."[16]

5

The *social institution* component makes routine those events that are seen as unique to the individual. Illness, fires, parenting, death, and so on are "controlled" by the actions and language of specialized professions who bury the dead, fight the fires, adjudicate marital distemper, and so on. Our personal tragedy in the loss of others through death is made routine by both rituals and rhetoric.

Kinship refers to basic relationships established by birth or marriage. Here biology meets culture and social structure to create and sustain the social system. Though the details of these relationships may vary by culture, place, and time, all social systems depend on such institutions. Kinship relationships contain, organize, and sustain basic emotional and practical dimensions of human nature in terms of courtship, marriage, reproduction, child care, inheritance property, and mourning rituals for the departed. In all of these and associated behavioral needs, the full network of those who have gone before, the elders of the kinship network, the grandparents, uncles, aunts, cousins of varying connectedness play a role in directing what is appropriate and inappropriate behavior for the members of this kinship network. The kinship network then serves the social system by institutionalizing the behaviors required for the management of sexual drives, providing future members of a society with nurture and care of infants (a species with relatively long period of early dependence) and primary socialization of children and youth, developing the social identity of individuals, ensuring a level of behavioral predictability in certain functional roles, and extending purpose to an entity larger than the individual through sharing a particular biosocial heritage and inheritance of this heritage in an orderly way.

The important social and emotional functions provided by kinship institutions ensure that there is a substantial supply of governmental, academic, and anecdotal data for the human ecologist. In the discussion on population as a social resource, we reported on the ecological consequences from fluctuations in reproduction from high numbers of people born compared to periods when smaller numbers are born. These fluctuations reflect larger socioeconomic pressures, but these are translated by kinship institutions that may transcend the larger reality where emphasis on producing more of our kin is valued when less might be more realistic. The demand on natural resources by increasing or decreasing populations is regulated by more than pure biological factors.

Another example of the influence on biophysical resources by kinship institutions is in inheritance of property institutions. A culture may assign landed property through primogeniture where the eldest son has primary right to the estate of his father. Or it may be regulated by entail where there is a necessary sequence as to the order of descent for the entailed property. If entail is the rule, then the ownership of farmland could become fragmented into economically inefficient units and compel further movement into higher elevation or wetland forests to compensate for the reduced family estate. Though primogeniture rules may seem "unfair," their biological impact may be advantageous under certain conditions. The point here is that kinship institutions are often a critical factor that must be considered in policy and planning efforts for human ecosystems. Further, there are substantial sources of data for identifying the connections affected by kinship institutions interacting with other ecosystem elements.

Health care institutions encompass the full range of organizations and activities that deal with the *health* needs of a human ecosystem. Health care in modern industrial societies is relatively complex, including primary care (personal and family health maintenance, outpatient activities by general practitioners), secondary care (such as services of specialists), and tertiary care (such as hospital procedures involving surgery).[17] Health care institutions are often measured by capacity (the number of doctors or hospitals per thousand people) or outcomes (infant mortality rates). In rural communities, primary care is often available locally; secondary and tertiary care is often provided on a regional basis. Hence, relatively small changes in the health institution (a doctor's retirement, the closing of a pharmacy) may have direct and indirect effects that ripple through the social system.

The collective problem of *justice* (law) faces all human social systems; its role in human ecosystems is critical. Two forms are central: distributive justice (who should get what, such as property rights) and corrective justice (how should formal norms be enforced, such as rules for punishment).[18] The legal system can be measured by both its practitioners (such as the number of lawyers or judges per thousand people) and its performance (number of trials or convictions). The contemporary legal system plays an important role in ecosystem management—

the courts influencing distributive justice through timber sale appeals and injunctions and meting punishment for resource crimes (such as poaching). Changes in legal institutions, such as new procedures for appeal or new laws (revision of the Endangered Species Act is an example), can dramatically and directly impact the use of natural resources, the development of capital, and other components of the human ecosystem.

To the human ecologist, *faith* (religion) as an institution has two components: its social function as a system of organizations and rituals that bind people together into social groups, and a coherent system of beliefs and myths.[19] Both are critical to human ecosystem functioning. Religion, like other social institutions, can be measured by diversity (range of religious practices), capacity (number of churches), or participation (percentage of the population claiming membership). Religion impacts the social system in many ways, altering social cycles (religious holidays), providing identity for both caste and clan, and influencing beliefs and myths. A change in faith (such as increased demands after a natural disaster) can have significant bearing on how effectively social systems adapt to new ecological and socioeconomic conditions.

The interplay of religion and ecology has multiple dimensions. J. Donald Hughes's extraordinary *Ecology in Ancient Civilizations* provides a historical foundation. E. O. Wilson's book-length essay *The Creation,* written as a letter to a Southern Baptist pastor, is an effort at religion/science consilience from a scientist's perspective. Pope Francis' recent encyclical letter *Laudato Si': On Care for our Common Home* reflects a nuanced understanding of environmental science (including "cultural ecology") and its relationship to Catholic theology and practice.[20]

All societies require a system for exchanging goods and services, and the institution of *commerce* (business/industry) is central to this exchange. Commerce includes not only the exchange medium but also the organizations that manage exchange, such as banks, markets, warehouses, retail outlets, and so forth. Modern industrialized societies (including their rural regions) rely on a mix of exchange styles; the typical U.S. rural community usually conducts its commerce through a mix of cash, credit, and barter. Commerce can be measured as capacity (such as the percentage of production capacity utilized, the number of banks) or as a flow (the number of transactions or the dollar value of a gross regional or local product). Commerce in rural areas, particularly in the west, depends largely on local natural resources (be it water, energy, timber, scenic value, or other values); a change in commerce can create a cascading set of impacts on other social institutions (such as sustenance), the social order (shifts in wealth or power), social cycles (such as recurring periods of economic recession), and critical resources (such as land or labor).[21]

Individuals are born into the world sorely lacking in the knowledge needed to survive, adapt, and interact with others. Hence, *education* (the transmission of knowledge) is a ubiquitous collective challenge: We must educate our young. While significant learning takes place in the home and on the streets, the educational institution largely functions through the school system, including public and private schools, teachers, school boards, and parent organizations.[22] Education can be measured in density (teacher/student ratios), input (dollars expended per student), and an output (percentage of high school seniors graduating). Changes in the educational system directly impact other components of the social system (such as the timing of leisure activities, the distribution of knowledge, the availability of skilled labor). Dramatic changes in the institution (such as school consolidation) can have significant effects on the entire human ecosystem.

Leisure (the culturally influenced ways we use our nonwork time) is an important institution in all but the harshest human ecosystems. Several studies suggest that industrialized societies have *less* leisure time per capita than agricultural or pastoral ones.[23] In industrialized societies, the recreation institution includes formally managed leisure opportunities (bowling alleys, wilderness areas, movie theaters, hunting and fishing spots) as well as informal pursuits (socializing, sexual behavior or courtship, resting) and specialized events (holidays, festivals, and so forth). Leisure can be measured as an amount (hours per day per capita), as a level of participation (percentage of adults with hunting permits), or as a range (number of festivals or special events). Changes in leisure can impact human ecosystems in several ways, as through direct impacts on commerce (a boom or bust in the tourist industry) and by changing social norms (a decline in festival attendance or a change in gender participation).

The political subsystem (*governance*) is at once a central component of human ecosystems and a result of numerous other components (such as organization, myths, legal institutions, and so forth).[24] Politics as an institution is a collective solution to the need for decision-making at scales larger than the clan or caste. It includes the modes of interaction between political units (such as states and counties), the processes of decision-making within political units (such as elections and legislative action), and the participation of citizens in political action (campaigns, party activity, referendum, and so forth). Government can be measured by its resources (tax receipts, authorized expenditures, employees) or its actions (laws passed, hearings held, violators punished, claims settled). Governments at several scales control critical natural resources, such as the ownership by cities, counties, states, and the federal government of forests, parks, reservoirs, canals, and other infrastructure. Hence changes in government action or process (for instance, revision to the Endangered Species Act) can have a significant influence on the human ecosystem.

The provision of *sustenance* (food, potable water, energy, shelter, and other critical resources) is a central and collective challenge facing all social systems. The management of that challenge and the production of necessary supplies requires agricultural and resource management institutions of some complexity; this is especially critical for resource-dependent communities such as timber towns, fishing villages, and park gateway communities.[25] Irrigation districts, farmer's cooperatives, timber companies, tree farm associations, extension offices, federal management agencies, and environmentally oriented interest groups are all components of the sustenance institution. Measures include organizational capacity (number of agents/ farm, acres in production), output (measured in dollar values or crop tonnage), and range of sustenance products (number of crops or timber types). As agriculture and resource management are the chief methods for transforming critical resources into necessary social system supplies, their importance to human ecosystem functioning is key. Changes in production, efficiency, or distribution can have effects throughout human ecosystems.

Security is a constant need to be reinforced by social institutions. As an infant, we need the security of our mother and, later, other relatives around us. As a community or a nation-state, we need food security, social security, health security, and security from threats of violence. We make these challenges routine with socialized rules about parenting taught and reinforced by peers and relatives and social service agencies to rectify errors on the edge of these communication loops. We need to have a level of trust in relatives and care-givers in our relationships where we believe in the other to reciprocate our mutual support within the security of a long-term relationship. Trust in the government or work organization makes possible the security of an anticipated pension or social security benefits or food stamps and other solutions to aging and survival. When politicians or corporate executives say that they can no longer honor that trust, then our sense of security is threatened. Without that trust and its institutions of security, our performance as workers, citizens, care-givers, and supporters of those in our relationships is less enthusiastic and often counterproductive.

Trust is not always forthcoming. Security includes armed institutions with missions of defense, invasion, occupation, and suppression of civil unrest. Out of 192 countries in the world, 163 currently maintain regular armed forces, and the number and diversity of militias, insurgencies, revolutionary groups, and terrorist organizations has increased dramatically since the end of World War II. The human ecological impact of warfare (war preparations, violent conflict, and postwar restoration) is significant, and "warfare ecology" is emerging as an applied subfield of human ecology.[26]

6

Human ecosystems are driven by cycles of change in the nonhuman world and by changes in human demography. A human population can fluctuate in terms of increases through births and deaths and immigration and emigration. Bulges at either end of these patterns challenge existing institutions. Climate change, extreme weather, and fluctuation in drought or wetness place demands on the existing social structures. We consider four of these timing cycles.

Humans have evolved with a series of *physiological cycles* that deeply influence social behavior. For example, diurnal cycles of night and day create peaks of labor and leisure activity; menstrual cycles control reproduction patterns. The life cycle is roughly similar across cultures: birth, childhood, labor, marriage, child-rearing, retirement, and death. Each stage of the life cycle creates expectations and norms for behavior (including the use of resources).[27] Measurement can include the proportion of the population at each stage of the life cycle. These cycles create predictable patterns of activity within the human ecosystem: park going during daylight hours, increases in energy demands during early morning hours (for showering, cooking, heating, and so forth), rituals at each juncture of life cycle stages (graduations, weddings, funerals). While physiological cycles may rarely change, they substantively impact human ecosystem functioning at several scales.

Beyond physiological cycles, *individuals* may follow time cycles that are personal and idiosyncratic. Examples are graveyard shifts for certain workers (bakers or police), part-time or seasonal work (such as agricultural field labor or lumbering), and personal patterns of recreation activity (weekend hiking or camping). These cycles impact social institutions and the use of natural resources. They can be measured by such indicators as employment patterns (for example, the proportion of part-time to full time workers). Changes in individual cycles can reflect alterations in labor needs, social institutions, or hierarchies of wealth.

Each of the social *institutions* described above have (or create) social cycles that control the flow of relevant activities.[28] The legal institution, for example, creates court seasons and trial days; the recreation and sustenance institutions set hunting and fishing seasons. These institutional cycles are critical to human ecosystem functioning, for they provide guidance and predictability to the ebb and flow of human action. Institutional cycles can be measured in terms of frequency (the number of times persons or groups participate), duration (the length of a hunting season), proportion (the percentage of the population involved), or intensity (the depth of the meaning assigned to the cycle, such as the funeral of a national leader). Changes in institutional cycles may directly impact the use of natural resources (for example, a year-round school calendar diversifying park-going patterns) and, importantly, the conduct of commerce (such as fishing seasons, field-burning periods, or fiscal year cycles of funding).

Not all cycles are socially constructed: *Environmental cycles* are natural patterns that can significantly influence the human ecosystem.[29] Environmental cycles include (but are not limited to) seasons, drought periods, El Niño patterns, biogeochemical cycles, and long-term climatological change. Drought cycles in the western United States, for example, impact natural resources such as wildlife and forests; the capital needs for dams, reservoirs, and other storage devices; agricultural institutions; litigation over water rights; and many other components of the human ecosystem. The cycles can be measured by intensity (such as global ambient air temperature), duration (length of growing season), or occurrence (the proportion of years in a decade with low precipitation). Changes in environmental cycles, such as the end of a drought, the movement of the seasons, or climate change can alter ecosystem and social system responses, often significantly.

7

The social order is the "mixer" of resources, institutions, and cycles. It must create and sustain some form of order in the various transactions driven by these other components. The expectation of certain levels of behavior and the understanding of one's role in various social transactions give an economy of understanding to guide our actions. We do not expect the bank clerk to pull a weapon on us to gain a deposit, and they do not expect us to use a weapon for our withdrawals. There are forms to fill out and procedures to follow. Our organized division of labor makes routine and predictable many of our daily encounters.

One of the key ways that social systems maintain coherence and the ability to function is through the use of *identity*. In sociological terms, identity is often ascriptive—it is assigned by society based on birth or circumstances rather than through the individual's actions or achievements. Caste or race, for example, is ascriptive: One is born into a racial category that then follows the individual throughout the life course. These identities are used (often through stereotyping or other generalizations) to differentiate people and manage interactions:

African-Americans claim affinity to one another by the ascription of race, Chinese make similar claims to each other by the same reckoning, both groups identify differences between them. Other identities are less ascriptive, such as class: Individuals can alter their class through changes in wealth, education, and occupation.

Several forms of identity are critical to human ecosystems. *Age* is important, for much of human activity is age-dependent: Certain occupations (such as mining) are mainly for the young; certain recreation activities (such as white water rafting) are likewise often specialized by age. *Gender* (the socially constructed masculine and feminine roles) is important both for its crucial impact on social norms and for its differential effects on social institutions—women and men having different access to capital, health care, wealth, power, and other features of the social systems. *Class* is important, though its definition is problematic.[30] Some social scientists define class in purely economic terms (based on occupation or income); others include sociocultural concerns (such as education or social norms). *Caste* is a socially created taxonomy for defining levels of social "cleanliness" or "uncleanness" in relations with other groups. It is usually associated with certain kinds of work. Like race, one is born into a caste and must follow the rules of that caste. Finally, *clan* (the extended family or tribal group) is crucial, both as a predictor of interaction (much recreation, for example, takes place with family members) and as a source of support. Clans routinely provide health care, financial assistance, even biophysical resources (such as food or other supplies) to members in need.

These identities can be measured in terms of diversity (the range of ethnic or age groups in a community) and/or distribution (the proportion of non-Caucasians within a population, the ratio of working-age individuals to dependents). Changes in identity usually impact social systems through an alteration in social norms; an influx of young people, Jews, women, and blue-collar workers leads to shifts in what is expected as well as in what people do; these shifts further alter the human ecosystem.

Social norms are rules for behavior, what Nicholas Abercrombie and his colleagues call the "guidelines for social action."[31] *Informal norms* are administered through community or social group disapproval: Deviating from the norm is noticed, but sanctions are slight. Speaking too loud in a museum or cheering not loudly enough at a football game are examples, as are norms for behavior in campgrounds, along trails, or on fishing boats. The full range of etiquettes for eating, socializing, courtship, and so forth are also informal norms. *Formal norms* are more serious and institutionalized; formal norms are usually codified in laws that not only prohibit certain actions but proscribe punishments for breaking such norms. Misdemeanor and felony laws are examples. Sometimes a community's informal norms may conflict with its formal (legal) norms. The results are "folk crimes," that is, activities that are against the law but not considered harmful by the population. Some kinds of wildlife poaching or illegal wood cutting can be classified as folk crimes.[32]

Norms can be measured by both their adherence (the proportion of a population following a social convention, such as marriage before childbirth) and/or deviance (the number of felonies per capita). Changes in social norms can impact social institutions (divorce directly impacts health and justice for women) and alter resource use.

An important mechanism for social differentiation, and for managing the social order, is *hierarchy*. In almost all social systems hierarchy is ubiquitous; inequality of access to scarce resources is a consistent fact across communities, regions, nations, and civilizations. Five sociocultural hierarchies seem critical to ecosystem functioning: wealth, power, status, knowledge, and territory.

Wealth is access to material resources, in the form of natural resources, capital (money), and credit. The distribution of wealth is a central feature of social inequality and has human ecosystem impacts; the rich have more life opportunities than the poor. *Power* is the ability to alter the behavior of others, either by coercion or deference. The powerful (often elites with political or economic power) can have access to resources denied the powerless; an example is politicians who make land use decisions and personally profit from these decisions at the expense of other citizens. *Status* is access to honor and prestige; it is the relative position of an individual (or group) on an informal hierarchy of social worth.[33] Cultures may vary as to what classes are granted high status (for instance, teachers are given high status in China, modest status in the United States). Status is distributed unequally, even within small communities, and high-status individuals (such as ministers or tribal elders) do not necessarily have access to wealth or power.

Knowledge is access to specialized information (technical, scientific, religious, and so forth); not all within a social system have such access. Knowledge provides advantages in terms of access to critical resources and the services of social institutions. Finally, *territory* is access to property rights (such as land tenure and water rights). Hierarchies of territory are created when some have strong land tenure (large tracts with secure ownership) and others weak tenure or are landless. This can vary by region. For example, in the arid U.S. West, water rights (granted by historical priority) may be especially crucial, as it is water that limits development.[34]

These critical hierarchies can be measured in several ways. Wealth can be measured by indicators such as the range of incomes or the proportion of the population that is below the poverty line. The distribution of power can be indirectly measured by certain decision-making activities, such as elections. It can also be measured by levels of domination and subordination—the disproportion of blacks and Latinos in prison or on death row, the "glass ceiling" faced by women workers, the persistence of spouse abuse, and the relationship between timber workers and company executives. Status can be measured by public polling techniques that capture public opinion; knowledge can be indicated by educational attainment. Territory can be measured by ownership patterns, the distribution of land by size (the proportion of landholders with large tracts), or the distribution of water rights (by acre/feet). Changes in hierarchies, by altering who has access to critical resources and social institutions, can dramatically alter the human ecosystem.

Social species have much of their spatial behavior directed by territorial defense and expansion.[35] People fight wars, organize police forces, and use terrorism to defend and expand property rights and territorial hegemony. The battles between Israel and Palestine are basically about contested claims over territory. These claims are not unlike other colonial battles such as the expansion of the United States through the expropriation of native people's lands or the expansion of Russian hegemony over control of land in satellite states. At the local level gang wars such as that between the Bloods and the Crips in U.S. cities are battles about who controls drug distribution or other economic activities at the local scale.

E. O. Wilson clearly identifies the characteristics and importance of territorial behavior for social species. He notes: "A majority of biologists . . . define territory by the mechanism through which the exclusiveness is maintained, without reference to its function. . . . I am convinced that this time the majority is right for practical reasons, that defense must be the diagnostic feature of territoriality. More precisely, territory should be defined as an area occupied more or less exclusively by animals or groups of animals by means of repulsion through overt aggression or advertisement. We know that the defense varies gradually among species from immediate aggressive exclusion of intruders to the subtler use of chemical signposts unaccompanied by threats or attacks."[36]

8

Within the basic structure of human ecosystems, there are also several key *flows*. It is these flows between the components that create the dynamics of human ecosystems. Six seem essential. The first is the flow of *individuals,* which may be flora and/or fauna of varying species, including *Homo sapiens*. The second is *energy,* in a variety of forms from mechanical to heat energy. The third flow is that of *nutrients* needed by flora, fauna, and persons, including food. A fourth key flow is that of *materials*. In this usage, materials include biophysical resources such as water and wood and human-made materials such as concrete and cocaine. The fifth flow is the flow of *information,* which can range from genetic information to cultural information and has been significantly affected by technology and the use of social media in recent years. The sixth flow is the flow of *capital,* in money form or its bartering equivalents.

These flows—within human ecosystems and between them as well—vary by rate, intensity, duration, frequency, and distribution. Flows between the structural components of human ecosystems designate most biophysical and sociocultural processes. Fighting a war, for example, requires immense and intensely targeted flows of individuals (invading armies), energy (stored and released in munitions), nutrients (both in base camps and on the battlefield whether land, sea, or air and able to withstand high or low temperatures and eaten with limited or no preparation), materials (in the form of military equipment), information (including propaganda and covert intelligence), and capital (to support all of the previous flows). Flows "activate" human ecosystems, generate a wide range of dynamic processes and patterns, and constitute the linkages between variables in the HEM. Hence, a realistic understanding of key flows is essential to understanding the structure and dynamics of human ecosystems.

The model in figure 6.1 denotes the complexity of human ecosystem patterns by labeling each linking arrow with the full range of potential flows. The model provides for a series of related flows, such as information flowing to the governance institutions, wealth flowing to individuals through the funding cycles of institutional banking, energy being converted to capital through the process of manufacturing, and the institutions of commerce, and so forth. These flows "bind together" human ecosystems in relative and varying coherence and have essential biophysical and sociocultural implications.

The parts and processes of human ecosystems are intricately linked through both positive (amplifying) and negative (dampening) feedback loops. Positive feedback is not necessarily good or advantageous; negative feedback is not necessarily bad or disadvantageous. For example, low water quality can increase disease, such as cholera, which, when spread through a population, can positively impact (that is, increase) water contamination in a dangerous cycle of declining water quality and health. Negative feedback loops can be advantageous in human ecosystems: Examples include soil microbes that transform bacterial waste, social laws and customs that constrain consumption, and so forth. Many processes and flows within human ecosystems exhibit both amplifying and dampening feedbacks. An examination of modern economic patterns (such as contemporary tax structures in industrial cities) reveals a complex mixture of subsidies, personal reciprocities, perverse and progressive incentives, and regulatory policy that make up postmodern economies and modulate key flows within these urban systems.

9

To conclude our detailed description of the components and variables in the HEM, we recognize that each element of a human ecosystem (with its components and variables) has the potential to influence the behavior of other elements in the system. The influence depends on many interactions, of course, and on the relative coherence of the components and variables. The influence of one portion of the system on another occurs in cascading first-, second-, and third-order effects.

This "cascade of consequences" is an essential implication of the model and its utility to resource management and solving environmental problems. Figure 6.2 illustrates variations in order effects. For example, changes in the flow of *energy* such as an embargo and resultant rationing may alter hierarchies of *power*, as those with access to fuel supplies gain influence. The resulting alteration of hierarchy may modify *norms* for behavior as those without fuel acquiesce to demands by the newly empowered "fuel barons." Changes in energy resources will also affect the *labor* and *technology* needed. Social institutions of *commerce* and *sustenance* will undergo adaptations, and so on. Flows of *materials, nutrients,* and *capital* will also adjust. Another example is when newly achieved access to drinking water in rural Africa ameliorates the need for lengthy travel to water supplies, freeing women to allocate time to education, and creating opportunities for economic development on the level of the village.

Such cascading effects are essential for understanding past and present ecological dilemmas. In the Ehrlichs' *One with Nineveh*, one explanation for the collapse of Mayan civilization elegantly illustrates historical third-order effects:

> It now appears that the [Mayan] civilization's demise was caused by a tangled complex of which the stress of overpopulation on limited agricultural land was a key component . . . early Mayan terraced fields, on which crops of corn, squash, gourds, and palms had been successfully grown for hundreds of years, had been buried under nearly a meter of clay sediment. The clay could not be farmed because it is rock-hard

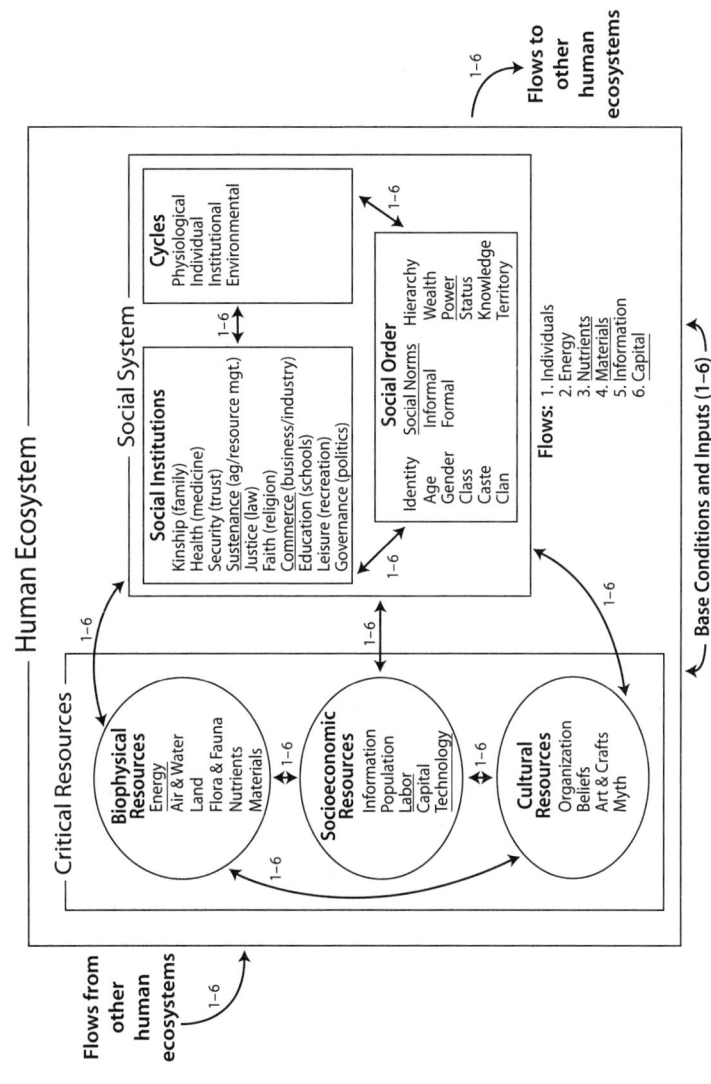

Figure 6.2. The Human Ecosystem Model illustrating "cascading effects" between components and variables

in the dry season and inundated in the wet [seasons]. . . . The sediment came from the erosion of denuded hillsides. Why were the hillsides denuded? Archaeologist Richard Hansen has assembled data indicating that, in addition to the gathering of wood needed for household fuel and construction, the deforestation of the hills was the result of increased demand for firewood for the kilns of the Mayan lime industry . . . [firewood demand] contributed substantially to the deforestation that led to erosion; erosion to clay sedimentation of farm fields; and clay sedimentation to hunger.[37]

Similar arguments have been made by Diamond in his detailed volume *Collapse*.[38]

Primary agency (another possible term is "driving force") within human ecosystems is the striving to meet human needs and wants at the scale of persons, social groups, and organizations. Bennett describes the practical importance of understanding human striving as an ecological driving force. He notes, "It is my conviction that a policy-oriented ecology will be achieved by focusing on strategic action, on the behavior of humans in dealing with nature and society in order to achieve their objectives, on ways of controlling such behavior, and on the adaptive or maladaptive effects of the behavior on the natural and social environments. What do people want? How do they get it? What are the consequences?"[39]

Bennett's trio of questions is both a repeatable mantra for human ecological research and the core drivers of environmental resource management policy: What do people want? How do they get it? What are the consequences? In our model, efficacy in achieving "advantageous adaptation" is based on social power, broadly defined to include social, cultural, political, economic, and military forms. Human ecosystems are hierarchically nested within human ecosystems at different scales. They are linked together through a skein of shared base conditions, interconnected components, cascading resource flows, adaptive mechanisms, and the primary agency of human needs and wants.

The evolution of human ecosystems occurs as structural revisions to a specific component or variables within that element. These revisions include alteration of the structure and/or influence of key components or variables, a shifting in relative coherence among components, or a change in the flux of resource flows. For example, new forms of energy conversion (such as the coal-stoked steam engine or fracking) alter the geography of commerce—with cascading consequences for everything from flora and fauna to hierarchies of political power. There is little evidence of directive force within human ecosystems (the myth of inevitable "progress"), even though some might wish it so. There *are* cumulative consequences of repeated and immeasurable adaptive actions by persons, groups, institutions, and nation-states.

Throughout this book we have indicated how the flows of individuals, energy, nutrients, materials, information, and capital influence the characteristics, stability, and change in the human ecosystem. These are the drivers of the adaptive changes necessary for the system to persist. Sometimes the change is in a similar frame, other times it requires major changes. Process is the eternal reality for any human ecosystem. The next several chapters present several applications of the model to human ecosystems.[40]

S • E • V • E • N

Goals, Strategies, and Tactics for Inquiry and Action

1

Our version of human ecology is a practical ecology, born out of the pressing need to solve problems. The application of the human ecosystem model as an organizing concept for resource management, environmental decision-making, and human ecological research is both real-world and worldly. Ecosystem theory is to be tempered by the complexity of practice.

In the preceding chapters, we have presented the evolution of biophysical-based ecological theories, concepts, and models not as analogies but integrated with sociocultural systems into an organic and dynamic whole: the HEM. Our model can help students, scientists, researchers, planners, policy-makers, and communities better understand the elements and interactions in the human ecosystem, and they can then use the HEM to develop strategies and tactics for a wide variety of complex problems in the coming decades.

Our inspiration comes from the work of Goethe, who was both a scientist and a poet.[1] He looked at humanity in all its faults. He admired our ability to err and yet to redeem ourselves because we continue to strive to experience and understand more. The challenge of combining art and science brings us closer to Goethe's ideal. We emphasize that the HEM is a menu of possibility organized by a logical structure that permits us to learn how the components and variables of the human ecosystem respond to particular forces of environmental change. We learn from our errors and strive to avoid these errors. The HEM encourages us that there is a means to strive and to try again—using both science and the creative arts.

Strategies are plans that combine science and art to move toward a goal. The *goal* to survive and ensure reproductive success is the basic biological interest of all species, and the HEM can help us develop strategies for a wide variety of ancillary goals for the quality and sustainability of our environment, the lives of people, and the stability of societies today. A watershed management plan—whether for the Mekong River in Southeast Asia, a tributary of the Mississippi River in the central United States, or a small stream in a valley in the Andes—may have a variety of goals. The goal may be to achieve a sustainable set of ecosystems that provide a wide variety of goods and services to the humans, flora, and other fauna that live in the watershed across diverse land uses, multiple ownerships, and many communities. This goal is quite different from the goal to manage an industrial waste stream for a food processing plant in Hanoi, Vietnam, to reduce pollution and create additional by-products for the market or the goal to manage medical procedures and increase children's health in a rural hospital in Botswana. *Goals* are the "valued achievements" held by cultures, nations, interest groups, households, and persons. Goals emanating from moral sources are colored by knowledge and prejudice (and the thin distinction between the two) and are ultimately based on the religious or secular faith of individuals and the missions of organizations.

Importantly, our model does not provide a ready source of goals. Goals are established by a particular social system in a specific location within a specific time period. HEM users must establish the appropriate goals for the situations they are facing. Our interest is to concentrate on providing HEM users working at many spatial and temporal scales with a model that helps them develop better strategies and tactics for their research, policy, planning, and management to achieve their goals.

2

Strategies are overarching plans or approaches to achieving particular goals or groups of goals. Strategies are broad visions or guides for effective action to resolve specific problems. During the initial stages of examining a complex problem involving biophysical and sociocultural systems and challenges, the HEM can be of direct use in "brainstorming" or using other group processes to list possible routes of action and paths of achieving what is valued. Potential consequences, anticipated consequences, and, as much as possible, "black swans" (rare, unexpected events that would have significant consequences if they happen) should all be explored.

Tactics are basic, practical tools for inquiry and the specific actions we take individually or collectively to implement a strategy. The tactic may be a change in the regulations of industrial chemical waste into water bodies, new safety measures for fracking activities, external chemical treatments to eliminate an invasive species within protected areas, new protocols in the emergency room of a hospital, a surge in counterinsurgency troops, and so forth. For example, generals may have strategic visions of large numbers of soldiers swooping behind enemy lines, but on the ground there are commanders worrying about the selection of the instruments of warfare needed, the skills of the soldiers, and the food, shelter, and medical needs that will be required to carry out the strategy. Strategies give us the direction; tactics suggest the means for getting there.

In research, we move from theoretical ideas about the consequences of certain events, their burdens and benefits, their distribution and sustainability, and the appropriate techniques of measurement of an array of optional actions that test the theories about the consequences of specific decisions to adopt one action or another. How might the consequences of "cap and trade" of pollution emissions balance with straight regulation at the source of pollution? The tactics we use to achieve the broader strategic direction often result in policies that become large-scale social experiments that require monitoring and evaluation if we are to avoid worst-case scenarios. Thus, a flexible and parsimonious conceptual framework is required to organize the tactics that impart a high degree of confidence in the prescribed course of action.

Our model (and this book) can be useful in evaluating the efficacy and efficiency of particular tactics, assessing the tradeoffs between alternative tactics, and identifying old tactics to solve new problems as well as designing new tactics appropriate to action. *Methods* are the specific techniques and actions for implementing a tactic—a neighborhood meeting to get information and concerns from those most affected by a new land use plan, a regional social science survey of a population that is part of a research project on community adaptation to decreased water availability, a grassroots campaign, or a national initiative. To select a specific tactic and method for a specific situation, we suggest that the reader consult the vast literature and handbooks on social science methods.

The HEM gives users a better chance to build a learning curve so that each event is not a unique surprise. We can gain from a legacy of similar events. Such knowledge is cumulative and can more effectively guide actions for future events, whether brought about by human or nonhuman events. The person seeking to understand the "tipping point" of human ecosystems can find excellent reports on goals and a huge library of social and biological science books on methods for doing research. Within this context, between the grand and the mundane of life, our model tries to keep to the middle, pragmatic range of thought and action—understanding the consequences of certain strategies and tactics.

3

All persons and all organized social units from the individual to the household, community, corporation, and nation-state have established theories and tactics or, more accurately, a logic of inquiry for collecting, sorting, and interpreting the information that can guide their decisions. Of course, these tactics vary considerably between decision-makers and the nature of the issue they are trying to resolve. Further, all such decisions have their own structured means for judging the validity, reliability, and ability to replicate the information being collected, interpreted, and applied. In one case it may be community tradition, in others a family vacation ritual, or market trends, or past performances, or a peer review that helps to stabilize and lend confidence to the methods by which decisions are made.

Our need to understand the structure and processes of a whole system rather than just selecting a few paired variables to stand for the whole system compels us to accept complexity as our new reality. Consider the wide array of actors who might be using the HEM to address a complex problem, such as an urban transportation system for a major city. A regional planner may be seeking the best mix of transportation types to serve a rapidly growing (or declining) metropolitan area. A graduate student in geography may be doing doctoral research that examines the probable consequences for different demographic groups using the transportation system. Farmers with small amounts of farmland near the urban area may want to know what the system of transportation chosen by the regional planners will mean for their weekly farmers market. Mayors of the city and nearby towns may wonder what the plan will mean for their future elections. Biologists want to know what it means for the habitats of endangered species. And this complexity may just be the start of consequent challenges. However, our goal remains—to ensure that all of these persons and many more can draw on the HEM to improve their present and future decisions.

Of course, for researchers, the standard for judging research quality remains the strategies and tactics of science. Still, constraints of time, resources, and complexity may limit the full application of this standard. We cannot stress too greatly that for many decisions there are several legitimate, though different, techniques of research and measurement than that of "science-based" research. Though many of these approaches are "science-like" in form and attitude, they are not at the hard-edged standard necessary for many science information collecting efforts.

Robert Merton reminded us that the traditional ideal of research, though logical, often fails to "describe much of what actually occurs in fruitful investigation. It presents a set of logical norms, not a description of the research experience. And, as logicians are well aware, in purifying the experience, the logical model may also distort it. Like other models, it abstracts from the temporal sequence of events. It exaggerates the creative role of explicit theory just as it minimizes the creative role of observation. For research is not merely logic tempered with observation. It has its psychological as well as its logical dimensions, although one would scarcely suspect this from the logically rigorous sequence in which research is usually reported."[2]

Merton's critique fits most of the decisions we make. A household decision about a new home, a new car, a new baby is based on some theory, legend, or myth shared in the household as to what the relevant needs, benefits, and constraints are. Yet, in the moment we often decide on factors other than the guiding rationality. Once the decision has been made, we then develop a rationale as to why we made the decision in the way that we did. We most often refer to our customary methods that guided our decisions: We went on the internet and studied the reviews of experts, we examined the product or service we were considering, we did a trial run of the product or service, and we interviewed friends and neighbors with knowledge of the situation.

The links between "strategies" and "tactics" and between "theory" and "method" are tested by fitting the decision into the potential array of HEM variables that encourage the decision-maker(s) to make choices in terms of questions that compel consideration of what has been left out or what are the most critical and less critical connections involved. When the questions are derived from "theory," the decision-maker is compelled to review whether the appropriate range of necessary and sufficient variables and their interactions has been considered for a given issue. The question is: "Have I considered these variables and how they interact with these other variables in the human ecosystem?" Further, each HEM variable has functional indicators and several appropriate techniques of measurement that tell us how certain actions influence the individual variables and the entire human ecosystem. The HEM shifts this linking between theoretical questions and key variables to a discourse of possible outcomes that can challenge the bias of design in many human ecosystem analyses. The HEM has the flexibility to encourage the researcher to consider transactions between critical variables and to combine quantitative and qualitative data sets. It does this not by following tendencies to reduce to one or two basic mechanistic variables but by expanding the array of possible variables to be considered for a given issue.

We move now to some specific examples where the HEM can provide new perspectives on the problem at hand. We begin with a re-analysis of a prescient article from the 1950s. A current HEM-based analysis of past human ecosystem challenges and case studies can help us avoid or, at least, be better prepared for and possibly mitigate or adapt to future difficulties.

4

The HEM has a logic of inquiry with the ability to monitor how our strategic designs and our tactical actions connect to the empirical world. In 1951 an article was published by W. F. Cottrell called "Death by Dieselization: A Case Study in the Reaction to Technological Change."[3] Cottrell was a practicing sociologist who was trying to understand factors that re-shape social structure and function and contribute to a body of sociological theory on such matters. He analyzed how technological changes intersect with social policy and issues of social equity.

It must be noted that this study was done in the heady post–World War II period of American global hegemony. The factories in the United States were serving the world; middle-class factory jobs were in the United States, not overseas. The Rust Belt had not emerged. Yet in this context Cottrell described a situation that has played itself out in many places around the world, including the salvaging of General Motors in 2009 in the United States. It is a predictive and a prescriptive analysis.

Cottrell explored the relationship between changes in a critical biophysical resource (energy technology) and a critical cultural resource (social organization) and, in turn, how these changes differentially affect attitudes and values (beliefs and social identity). His method was a case study of an "ideal type of railroad town" that is "not complicated by other extraneous economic factors." He collected his data from public documents such as the local newspaper, interviews with key informants of different social groups, and the logic of the system he was examining. The town was built as a necessary element of the railroad when steam was the primary energy converter: "Thus the location of Caliente [name given by Cottrell to the case study town], as far as the railroad was concerned, was a function of boiler temperature and pressure and the resultant service requirements of the locomotive."[4]

Changes in technology gained from World War II resulted in the introduction of a diesel locomotive, which required a completely different set of needs from that of a train run by steam. "It requires infrequent, highly skilled service, carried on within very close limits, in contrast to the frequent, crude adjustments required by the steam locomotive."[5] World War II pressed the speed of adoption of the new technology and was stimulated by government demands and subsidy. In short, the location of Caliente to support the train system with a steam engine was no longer ideal. The town had built social institutions such as schools (education), churches (faith), social clubs (leisure), commercial enterprises (commerce), and government and public services (governance) based on the tax returns of the railroad company. With the diesel locomotive now the core of the train system, these social institutions were irrelevant, and the structures soon began the process of being boarded up.

Cottrell ran through the groups that pay heavy costs for the shift in the prime energy converter from steam to diesel—demographic factors of older more skilled workers (age), families with children (kinship), employees with higher status authority and management (status), local merchants (commerce), bond holders (hierarchy), churches (faith), and town hall (governance). All paid the costs of the gains to producers and consumers of the goods carried by the railroad. The local people firmly believed in the "American" system in which the world is determined by "laws of supply and demand" and "progress is inevitable." Yet the reality was that those who, in good faith, held to these ideas were those who paid the highest cost of the change. As Cottrell notes the "'good citizens' who assumed family and community responsibility are the greatest losers."[6]

After presenting his analytic description of the social consequences of the shift in a key energy converter, Cottrell examines the emotional impact of these changes. "Defense of our traditional system of assessing the costs of technological change is made on the theory that the costs of such change are more than offset by the benefits to 'society as a whole.' However, it is difficult to show the people of Caliente just why *they* should pay for advances made to benefit others whom they have never known and who, in their judgment, have done nothing to justify such rewards.... They do not consider that the 'American Way' consists primarily of acceptance of the market as the final arbiter of their destiny."[7] Cottrell expects that the feeling of "injustice" would compel the local people to seek changes in the rules of the game (justice) through new rules by the government.

Cottrell's "autopsy" of Caliente's disaster is similar to the studies of community response after disasters such as heat waves, tornadoes, hurricanes, earthquakes, forest fires, and floods. The difference is that we do not often think of technological change that slowly kills a village in the same way as a "natural" disaster that may destroy an entire village in a few minutes or hours. Such drama brings reporters and media coverage and a sense of great peril even though all of these natural events are regular events in the tornado- and flood-prone areas. But in the case of Caliente it was events far beyond the control or understanding by the local people, who "did the right things as they were expected to do." The residents of the town were clustered together as victims of unknown powers every bit as devastating as an epidemic of swine flu. The contours of the event and the affected elements of the human ecosystem are similar in both pattern and process. But the worldwide public attention and demand for immediate solutions are not the same.

We encourage the reader to review this interesting article from six decades ago that seems so lively and germane to our experiences as we confront the current challenges of information technology and global economic flatlands, which place under pressure persons who have "followed the rules" and through no fault of their own are now pushed to bear great costs. The guiding data and observations for the study are clear and appropriate. The possible policy solutions flow from the analysis. The author clearly identifies the factors driving the causes and effects influencing the biosocial environment of a small single-industry town and the consequences for many elements of their biosocial human ecosystem. The lessons to be learned from this mid-twentieth-century case study are many and can provide guidance around the world as industrialization and technological changes impact today's human ecosystems.

Eric Klinenberg's in-depth "social autopsy" of the 1995 heat wave deaths in Chicago is another example of "overlooked" tragic events with lessons for decision-makers. He reports the high number of mortalities above the norm was almost casually attributed to "natural" factors or poverty and age by health officials and politicians. Yet "the proportional death toll from the heat wave in Chicago has no equal in the record of U.S. heat disasters." His careful analysis demonstrates a greater number of deaths than official reports and identifies the social variation within and between neighborhoods with high and low mortality. He notes, "This study establishes that the heat wave deaths represent what Paul Farmer calls 'biological reflections of social fault lines' for which we, not nature, are responsible."[8]

The decision-maker looking for means to reduce mortality from future heat waves would want to look beyond the easy explanation that there are poor and older persons "naturally" susceptible to such extreme events. Secondly, there is the need to understand the observed conditions associated with variation in mortality rates. Klinenberg looks at the larger urban context and then focuses on two adjacent residential areas with similar biosocial conditions, but great differences in death rates. He notes, "North Lawndale . . . experienced 19 heat related-deaths for a rate of 40 per 100,000 residents and South Lawndale (colloquially known as Little Village) . . . had 3 deaths and a rate of less than 4 per 100,000 residents—ten times fewer than North Lawndale." Yet, these areas had "similar microclimates, almost identical numbers of seniors living alone and seniors living in poverty. The community areas, then naturally controlled for the weather and the subpopulation of people thought to be most at risk of heat wave death."[9]

Both of these two areas were mostly composed of minority groups. However, their ethnic characteristics were different. North Lawndale was 95 percent black and South Lawndale was 85 percent Latino. Though important, Klinenberg does not accept this as primary "cause" for the observed variation in death rates. He notes, "Latinos in Little Village did not experience the particular constraints of ghettoization, the rapid and continuous abandonment of institutions and residents, or the arson and violence that contribute to the destruction of the local social ecology. The second reason is that . . . the area has become a magnet for Mexican and Central American migrants. The continuous migration of Mexican Americans to this community area has replenished its human resources and regenerated the commercial economy of retailers and small local businesses . . . while North Lawndale lost more than half of its population between 1979 and 1990, Little Village grew by roughly 30 percent. . . . There are only a handful of abandoned buildings and empty lots in the area."[10]

The story is that a steady flow of immigrant populations from different cultural areas, but sharing a similar religious and linguistic cultural heritage, replenished the human resources and kept the streets active and the many small enterprises vital and growing. The factors that sustained the post–World War II migration to the outer suburbs were improvements in the financial resources, the technical innovations of new communities like Levittown, and federal tax policies and subsidies that increased the opportunity for the middle class (black and white) to leave the urban problems behind. However, the urban population movements of the twentieth century did not have a steady inflow of immigrant populations that could replenish and "fill in" already-established institutional structures. So, for the depressed neighborhoods, unlike earlier migrations from the inner city, there was no process where the cohorts moving up in class position and residential areas had their places being filled by opportunistic newcomers. These depressed neighborhoods were default places occupied by populations who had nowhere else to go. As Klinenberg suggests, they were trapped in "ghettoized" places.

The HEM provides a format for combining findings from studies such as those done by Cottrell and Klinenberg into a set of predictive questions about similar events. Like medical autopsies, the goal is to learn from detailed analysis of a given case what the regularities are that led to a particular outcome. These "autopsies" direct future diagnosis of regularly occurring events, such as floods, hurricanes, and technological forces that transform human ecosystems. Better strategies and tactics for coping, recovering, restoring, and rebuilding human ecosystems can emerge from the model.

5

In the 1980s and early 1990s, anthropologists such as Moran, sociologists such as Burch and his colleagues and students, and ecologists such as H. T. Odum and E. P. Odum employed the human ecosystem as a theoretical framework. It was applied to archeological research, energy policy, threats to national parks, and anthropogenic impacts on biodiversity. Machlis and others produced an interactive atlas for the Puget Sound region to guide biodiversity conservation, and Machlis and R. G. Wright proposed a system of social indicators to complement biophysical monitoring in biosphere reserves.[11] Force and Machlis used the HEM as a foundation for teaching social impact assessment to over 180 mid-career international professionals from thirty-five developing countries in a six-week land use and community forestry workshop each summer from 1981 to 1994. Burch began work in Baltimore during this period, which will be described in detail in chapters 8 and 9.

Before we move to the Baltimore work, we present a case study from the Pacific Northwest in the United States that provided the opportunity for significant evolution of the early HEM to address twenty-first-century challenges.

6

In April 1993, U.S. President Bill Clinton held—and personally attended—a "Forest Summit" to discuss the situation in the Pacific Northwest and to address the question: "How can we revive the health of lands ravaged by logging and preserve the country's last virgin stands of temperate rain forests while rebuilding the devastated economies and divided communities?" The situation addressed by the Forest Summit was regionally known as the "spotted owl crisis" and had been the subject of controversy and turmoil in the timber-dependent towns of the Pacific Northwest for a decade. During the summer of 1993, a team of scientists and technical experts conducted an assessment of options to address the situation. This assessment provided the scientific basis for the Environmental Impact Statement and Record of Decision to amend Forest

Service and Bureau of Land Management planning documents within the range of the northern spotted owl.[12]

Three months after the April 1993 Forest Summit, President Clinton also directed the Forest Service to "develop a scientifically sound and ecosystem-based strategy for management of eastside forests." The project, known as the Interior Columbia Basin Ecosystem Management Project (ICBEMP) was initiated by the U.S. Department of Agriculture Forest Service and the U.S. Department of Interior Bureau of Land Management in late 1993.[13] In addition to numerous scientific studies that were conducted over the next two to four years, two draft Environmental Impact Statements (EIS) were produced in 1997, and the Final EIS and Proposed Decision was released in December 2000. Numerous protest comments were received and it was not until January 2008 that a Memorandum of Understanding was signed by six federal agencies to cooperatively implement *The Interior Columbia Basin Strategy.*

Machlis and Force were initially commissioned to develop a system of social indicators that could establish a baseline to use in monitoring future management actions in the ICBEMP region. The work focused on the Upper Columbia River Basin, which included the entire state of Idaho and thirteen counties in western Montana. Later, the contract was expanded to use the same methods for the states of Oregon and Washington.

Although the HEM was being applied to a variety of situations, as stated previously, and through these applications and research projects was evolving from the models presented in the figures in chapter 3 of this book, significant enhancement and development of the model was made during work on the Upper Columbia River Basin. For example, "Universal Supply Elements" became "Critical Resources" with three components: "Natural Resources," "Socioeconomic Resources," and "Cultural Resources," and several variables were added to each of these components. "Social Institutions" was added as a component of the Human Social System. Based on the literature, interdisciplinary discussions with colleagues and students, as well as a commitment to a parsimonious but reasonably complete model, the HEM that framed the 1995 project is presented in figure 7.1.

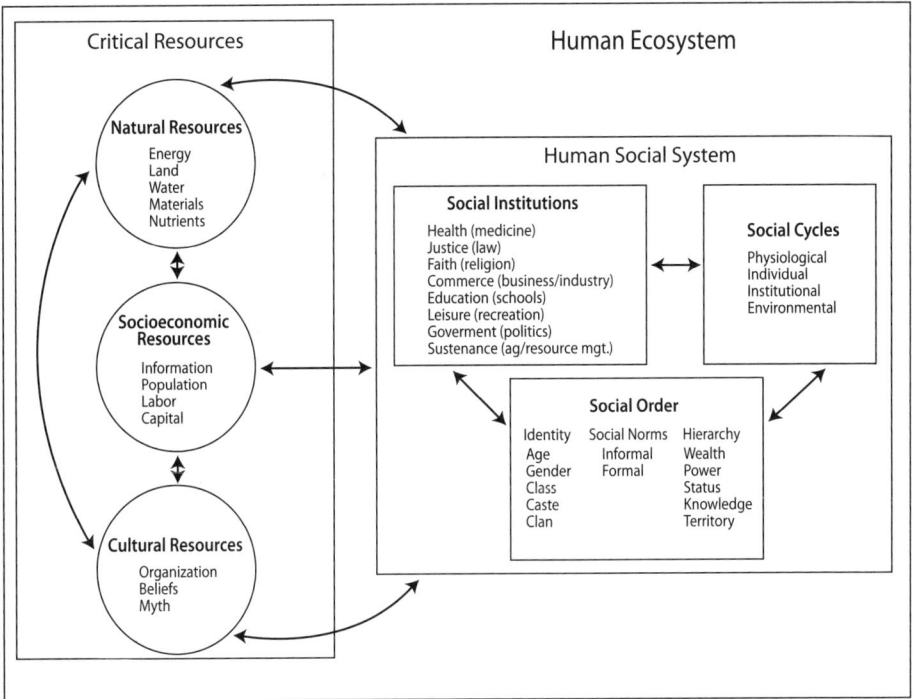

Figure 7.1. The Human Ecosystem Model: Equilibrium version (from Machlis et al., "The Human Ecosystem, Part I," 1997)

Given that the goal was to develop a system of social indicators that could establish a baseline to use in monitoring future management actions in the ICBEMP region, the initial strategy was to develop a monitoring system that focused on social indicators that could be easily understood and used by an interdisciplinary team of scientists and managers and were available from secondary, stable, public data sources. The rich and extensive literature on social indicators was explored— definitions, measures, and applications in a wide variety of situations and problems facing societies around the world. Some social indicators are based on simple theories, some on complex, barely understandable theories, and some appear to be a somewhat random collection of possible social measures with readily available data.

These activities led to the following definition of social indicators for ecosystem management in the Columbia River Basin project area: "an integrated set of social, economic, and ecological measures available to be collected over time and primarily derived from available data sources, grounded in theory and useful to ecosystem management and decision making."[14]

The definition has several important implications. Social indicators are more than a set of social, economic, and ecological measures—they are an *integrated* set. The measures for indicators are numerical values used to calculate the indicator, such as the percentage of the population that vote by a specific political party as a measure of the variable *beliefs* or the ratio of females to males as one measure of an indicator of the variable *gender*, which is an important form of identity. Social indicators are largely developed from existing data sources and are not dependent on primary data collection, which is often an expensive endeavor in terms of expertise, funds, and time. Social indicators should be available and consistent over time to be used in monitoring. Thus, they should allow for systematic comparisons across space and time if they are repeatedly collected. They are organized around an explicit theoretical framework that provides a rationale for selecting individual indicators and their measures. Indicators are multidisciplinary and reflect social, economic, and ecological concerns. Social indicators were sought that provide "usable knowledge" that is relevant for monitoring, decision-making, policy analysis, research, planning, and management.[15]

One of the first steps in using the HEM is to determine the boundaries of the problem and the unit of analysis to be used. This must be done whether using the HEM for a spatially defined application, such as most natural resource and environmentally based situations; a temporally defined application, such as developing a plan for a new policy to be implemented within a designated period; or an organizationally defined issue, such as a healthcare delivery system for a rural area. The boundary for the ICBEMP project was defined geographically by the river and then slightly reduced by the funds for the project. As stated above, the social indicators and HEM project covered Idaho, Oregon, Washington, and thirteen counties in western Montana.

After an initial investigation, we decided to use the county as the level of analysis. Readily available, high-quality secondary data at this scale are consistently collected at regular intervals for many of the variables in the HEM. The data can be easily compared across the United States, if needed. The county is a major unit of analysis for national census efforts and is a stable geographic unit for time-series data. In the United States, counties are also important administrative units for government regulations and policies, including planning and zoning commissions, which have significant impacts on land use within ecosystems. In the Pacific Northwest at the time of the ICBEMP project, county governments were expanding their environmental management capabilities. Nationally, in the 1980s, four of the top five issues facing county governments were environmental: solid waste, land use and zoning, water supply and sewage, and toxic waste.[16]

There is a wide variety of potential indicators for each variable in the HEM. In many cases, there are several appropriate measures for each indicator. For the ICBEMP project, the choice of indicators and measures for the HEM variables was based on several criteria: a relevance to ecosystem management activities; ease of understanding and interpretation by resource managers; availability of data at the county level; and accessible, high-quality data. The indicators and measures used for each of the HEM variables for Idaho and Montana are presented in table 7.1. In the final report to the ICBEMP project leaders, the data were presented in tables with the county name and the value of the measure for that indicator. The Idaho and Montana data were also presented in an atlas.[17] Table 7.2 includes the calculations that were used to develop the measures for the indicators for Oregon and Washington. Table 7.3 provides information about the data sources and organizations where secondary data were collected for the variables and indicators.

The HEM was well received by the project managers and of interest to many of the participants in workshops held during the project. The social indicator data collected using the HEM as a guide during the project is now part of the materials available on the Interior Columbia Basin Ecosystem Management Project website.[18]

Table 7.1. Social indicators used for the technical assessment of Idaho and Montana counties in the ICBEMP

Variable	Indicator	Measure (Date of Collection)
Natural Resources		
1. Energy	Occupied housing units heated with wood	% occupied housing units heated with wood (1990)
2. Land	Federal land	% land owned by federal government (1992)
	Population density on nonfederal land	Number of people per acre of nonfederal land (1990/1992)
3. Water	Not available	
4. Materials	Material production	Dominant manufacturing or extractive industry (1987)
5. Nutrients	Agricultural product	Ratio of cash value of crop products to livestock products (1992)
Socioeconomic Resources		
6. Information	Library loans	Number of books loaned by public libraries per capita per year (1993)
7. Population	Total population	Total resident population (1990)
	Rural population	% total population residing in rural areas (1990)
8. Labor	Unemployment	% civilian labor force unemployed (1989)
9. Capital	Bank deposits	Cash value of monthly bank deposits (June 1989)
	Income	Median household income (1989)

Variable	Indicator	Measure (Date of Collection)
Cultural Resources		
10. Organization	Not available	
11. Beliefs	Votes by political party	% votes cast for Republican presidential candidate (1992)
12. Myth	Major religion family	Major religious group (1990)
Social Institutions		
13. Health	Infant mortality	Number infant deaths per 1,000 live births (1988)
	Physicians	Number physicians per 100,000 population (1990)
14. Justice	Law enforcement	Number police officers with arrest powers per 1,000 population (1992/1990)
15. Faith	Religious adherents	% population claiming adherence to an established religion (1990)
16. Commerce	Earnings	Cash value of earnings in all industries (1988)
17. Education	High school graduates	% adult population graduated from high school (1990)
18. Leisure	Not available	
19. Government	Voting rate	% population >18 years of age participating in presidential elections (1992)
	Local government finances	Cash value local government expenditures per capita (1986–87/1990)
20. Sustenance	Resource-related employment	% labor force in agriculture, forestry, fisheries, mining (1990)
	Land use	Acres of irrigated land (1992)

(continued)

Table 7.1. (*Continued*)

Variable	Indicator	Measure (Date of Collection)
Identity (social order)		
21. Age	Median age	Median age (1990)
	Dependency ratio	% persons <18 and >64 years of age (1990)
22. Gender	Women in labor force	% adult women in labor force (1990)
	Sex ratio	Ratio of females to males (1990)
23. Class	Professional and skilled employment	% labor force in professional occupations (1990)
24. Caste	Ethnic/racial composition	% population in ethnic/racial groups (1990)
25. Clan	Household composition	% households of with children under 18 headed by single parents (1990)
Social Norms (social order)		
26. Formal	Crime	Number of serious crimes known to police per 100,000 population (1991)
27. Informal	Divorce rate	Number of divorces per 1,000 population (1987/1990)
Hierarchy (social order)		
28. Wealth	Poverty rate	% persons living below poverty level (1990)
29. Power	Elected positions	Number of elected positions per 1,000 population (1994/1990)
30. Status	Not available	
31. Knowledge	College graduates	% adult population who are college graduates (1990)
32. Territory	Homeownership	% housing units occupied by owner (1990)

Variable	Indicator	Measure (Date of Collection)
Social Cycles		
33. Physiological	Elderly population	% population >69 years of age (1990)
34. Individual	Employment terms	Full-time workers (1990)
	Work days	Seasonal workers (1990)
35. Institutional	Not available	
36. Environmental	Not available	

Note: At the time of the case study, *Air*, *Flora*, and *Fauna* were not in the model.
Source: J. E. Force and G. E. Machlis, "Human Ecosystem, Part II," 377–78.

Table 7.2. Social indicators, measures, and calculations used for the technical assessment of Oregon and Washington counties in the ICBEMP

Variable	Indicator	Measure (Date of Collection)	Calculation
Natural Resources			
1. Energy	Occupied housing units heated with wood	% occupied housing units heated with wood (1990)	Number of occupied housing units heated with wood *divided by* total number of occupied housing units
2. Land	Federal land	% land owned by federal government (1976: Oregon; 1983: Washington)	Number of acres owned by federal government *divided by* total number of acres in county
	Population density on nonfederal land	Number of people per acre of nonfederal land (1990/1976 or 1983)	Total population *divided by* number of acres of nonfederal land
3. Water	Not available		
4. Materials	Material production	The dominant manufacturing or extractive industry in county (1987)	
5. Nutrients	Agricultural product	Ratio of value of crop products to value of livestock products (1992)	Cash value of crop products *divided by* cash value of livestock products

Socioeconomic Resources

6. Information	Not available		
7. Population	Total resident population	Total resident population (1990)	
	Rural population	% population living in rural areas (1990)	Rural population *divided by* total population
8. Labor	Unemployment	Civilian labor force unemployment rate (1989)	
9. Capital	Bank deposits	Bank deposits (June) (1989)	
	Income	Median household income (1989)	

Cultural Resources

10. Organization	Not available		
11. Beliefs	Votes by political party	% votes cast for Republican presidential candidate (1992)	Number of votes cast for Republican candidate *divided by* total votes cast
12. Myth	Major religion family	% practicing major religions (1990)	

(*continued*)

Table 7.2. (*Continued*)

Variable	Indicator	Measure (Date of Collection)	Calculation
Social Institutions			
13. Health	Infant mortality	Number infant deaths per 1,000 live births (1988)	
	Physicians	Number physicians per 100,000 population (1990)	
14. Justice	Law enforcement	Number police officers with arrest powers per 1,000 population (1993/1990)	Number of police officers with arrest powers *divided by* number of total residents (convert to per 1,000 residents)
15. Faith	Religious adherents	% population claiming adherence to an established religion (1990)	
16. Commerce	Earnings	Cash earnings in all industries (1988)	
17. Education	High school graduates	% high school graduates among persons 25 or older (1990)	
18. Leisure	Not available		
19. Government	Voting rate	% population >18 years of age participating in presidential elections (1992)	Number of votes cast for president *divided by* total population >18 years of age

	Local government finances	Direct government expenditures per capita (1986–87/1990)	Direct expenditures *divided by* total population
20. Sustenance	Resource-related employment	% employed persons in agriculture, forestry, fisheries, mining (1990)	Number of employed persons in agriculture, forestry, fisheries, and mining *divided by* total civilian labor force
	Land use	Acres of irrigated land (1992)	
Identity (social order)			
21. Age	Median age	Median age (1990)	
	Dependency ratio	% persons <18 and >64 years of age (1990)	Total population minus persons 18–64 *divided by* total population
22. Gender	Women in labor force	% all women in labor force (1990)	Number of women working full time *divided by* all women in labor force, including those not currently employed
	Sex ratio	Ratio of females to males (1990)	Female population *divided by* male population
23. Class	Professional and skilled employment	% workers that are professional and skilled workers (1990)	Number of employed persons in health, education and other professions *divided by* total civilian labor force

(*continued*)

Table 7.2. (*Continued*)

Variable	Indicator	Measure (Date of Collection)	Calculation
24. Caste	Ethnic/racial composition	% population in ethnic/racial groups (1990)	Number of Black, American Indian, Asian, Hispanic and other races population *divided by* total population
25. Clan	Household composition	% households of single parents with children under 18 (1990)	Number of male householders, no spouse present, with own children plus number of female householders, no spouse present, with own children *divided by* All family households with persons under 18 years old
Social Norms (social order)			
26. Formal	Crime	Number of serious crimes known to police per 100,000 population (1991)	
27. Informal	Divorce rate	Number of divorces per 1,000 population (1987/1990)	

Hierarchy (social order)			
28. Wealth	Poverty rate	% persons living below poverty level (1990)	Number of persons below poverty level *divided by* persons for whom poverty status has been determined
29. Power	Elected positions	Number of elected positions per 1,000 population (1994/1990)	Number of elected positions *divided by* total population
30. Status	Not available		
31. Knowledge	College graduates	% adults who are college graduates (1990)	
32. Territory	Homeownership	% housing units occupied by owner (1990)	Number of housing units occupied by owner *divided by* total number of housing units
Social Cycles			
33. Physiological	Elderly population	% of population >69 years of age or older (1990)	Number of people >69 years of age *divided by* total population
34. Individual	Employment terms	Full-time workers (1990)	% full-time workers *divided by* all workers
	Work days	Seasonal workers (1990)	% seasonal workers *divided by* all workers
35. Institutional	Not available		
36. Environmental	Not available		

Source: Jo Ellen Force, David Fosdeck, and Gary E. Machlis, "Monitoring Social Indicators for Ecosystem Management: The Technical Assessment Data for Oregon and Washington," Interior Columbia Basin Ecosystem Management Project under Order #43-0E00-5-5269, 1995, pp. 11–14.

Table 7.3. Data sources for social indicators used in technical assessment for Oregon and Washington counties

Variable	Data Source	Organization
1. Energy—homes heated with wood	1990 Decennial Census (available on CD-ROM)	U.S. Department of Commerce, Bureau of the Census, Suitland, MD 20233
2.1 Land—Square miles per county	County and City Data Book, 1994	U.S. Department of Commerce, Economics and Statistics Administration, Bureau of the Census, Suitland, MD 20233: Government Printing Office, Washington, D.C.
2.2 Land—acres of federal land per county	OREGON: *Atlas of Oregon* WASHINGTON: *1994 Washington State Yearbook*	OREGON: Published by University of Oregon Books; available in University of Idaho Library WASHINGTON: Available at the University of Idaho Library
2.4 Land—acres nonfederal irrigated land	1992 Census of Agriculture, Geographic Area Series for ID and MT	U.S. Department of Commerce, Bureau of the Census, Suitland, MD 20233
4. Materials—dominant industry	1987 Census of Manufactures, Geographic Area Series for ID and MT	U.S. Department of Commerce, Bureau of the Census, Suitland, MD 20233
5. Nutrients—dominant product	1992 Census of Agriculture, Geographic Area Series for ID and MT	U.S. Department of Commerce, Bureau of the Census, Suitland, MD 20233
7. Population—resident population	1990 Decennial Census (available on CD-ROM)	U.S. Department of Commerce, Bureau of the Census, Suitland, MD 20233

Variable	Data Source	Organization
Rural population		
8. Labor—unemployment rate	USA Counties, CD-ROM, 1989	U.S. Department of Commerce, Bureau of the Census, Suitland, MD 20233
9.1 Capital—bank deposits	USA Counties, CD-ROM, 1989	U.S. Department of Commerce, Bureau of the Census, Suitland, MD 20233
9.2 Capital—median household income	1990 Decennial Census (available on CD-ROM)	U.S. Department of Commerce, Bureau of the Census, Suitland, MD 20233
11. Beliefs—% votes cast Republican	*County and City Data Book, 1994*	U.S. Department of Commerce, Economics and Statistics Administration, Bureau of the Census, Suitland, MD 20233: Government Printing Office, Washington, D.C.
12. Myth—major religion family	*Churches and Church Membership, 1990*	Glenmary Research Center, 750 Piedmont Ave., NE, Atlanta, GA 30308
13. Health—infant mortality	Both in *County and City Data Book, 1994*	U.S. Department of Commerce, Economics and Statistics Administration, Bureau of the Census, Suitland, MD 20233: Government Printing Office, Washington, D.C.
Number of physicians		
14. Justice—police officers	*Uniform Crime Reports: Crime in the United States (1993)*	U.S. Department of Justice, Federal Bureau of Investigation, Washington, D.C. 20535
15. Faith—religious adherents	*Churches and Church Membership, 1990*	Glenmary Research Center, 750 Piedmont Ave., NE, Atlanta, GA 30308

(continued)

Table 7.3. (*Continued*)

Variable	Data Source	Organization
16. Commerce—earnings in all industries	1990 Decennial Census (available on CD-ROM) (1988 data)	U.S. Department of Commerce, Bureau of the Census, Suitland, MD 20233
17. Education—high school graduates	*County and City Data Book, 1994*	U.S. Department of Commerce, Economics and Statistics Administration, Bureau of the Census, Suitland, MD 20233: Government Printing Office, Washington, D.C.
19.1 Government—Total votes cast for president	*County and City Data Book, 1994*	U.S. Department of Commerce, Economics and Statistics Administration, Bureau of the Census, Suitland, MD 20233: Government Printing Office, Washington, D.C.
Local government finances		
19.2 Government—Population > 18	1990 Decennial Census (available on CD-ROM)	U.S. Department of Commerce, Bureau of the Census, Suitland, MD 20233
20.1 Sustenance—% persons employed in forestry, agriculture, fishing, mining	1990 Decennial Census (available on CD-ROM)	U.S. Department of Commerce, Bureau of the Census, Suitland, MD 20233
20.2 Sustenance—Irrigated cropland	1992 Census of Agriculture, Geographic Area Series for ID and MT	U.S. Department of Commerce, Bureau of the Census, Suitland, MD 20233
21. Age—Median age	1990 Decennial Census (available on CD-ROM)	U.S. Department of Commerce, Bureau of the Census, Suitland, MD 20233

Variable	Data Source	Organization
Dependency ratio		
22. Gender—Women in the labor force	1990 Decennial Census (available on CD-ROM)	U.S. Department of Commerce, Bureau of the Census, Suitland, MD 20233
Sex ratio		
23. Class—Professional and skilled employment	1990 Decennial Census (available on CD-ROM) (1990 data)	U.S. Department of Commerce, Bureau of the Census, Suitland, MD 20233
24. Caste—Ethnic/racial composition	1990 Decennial Census (available on CD-ROM)	U.S. Department of Commerce, Bureau of the Census, Suitland, MD 20233
25. Clan—Household composition	1990 Decennial Census (available on CD-ROM)	U.S. Department of Commerce, Bureau of the Census, Suitland, MD 20233
26. Formal norms—serious crimes known to police	*County and City Data Book, 1994*	U.S. Department of Commerce, Economics and Statistics Administration, Bureau of the Census, Suitland, MD 20233: Government Printing Office, Washington, D.C.
27. Informal norms—divorce rate	*Vital Statistics of the United States, 1987, Vol. 3: Marriage and Divorce*	U.S. Department of Health and Human Services
28. Wealth—poverty rate	1990 Decennial Census (available on CD-ROM)	U.S. Department of Commerce, Bureau of the Census, Suitland, MD 20233
29. Power—elected positions	OREGON: *Oregon Blue Book* 1987–88, Oregon Secretary of State Office	Available at the University of Idaho Library
	WASHINGTON: *1994 Washington State Yearbook*	Available at the University of Idaho Library

(*continued*)

Table 7.3. (*Continued*)

Variable	Data Source	Organization
31. Knowledge—college graduates	*County and City Data Book, 1994*	U.S. Department of Commerce, Economics and Statistics Administration, Bureau of the Census, Suitland, MD 20233: Government Printing Office, Washington, D.C.
32. Territory—owner-occupied housing	1990 Decennial Census (available on CD-ROM)	U.S. Department of Commerce, Bureau of the Census, Suitland, MD 20233
33. Physiological—elderly population	1990 Decennial Census (available on CD-ROM)	U.S. Department of Commerce, Bureau of the Census, Suitland, MD 20233
34. Individual—full time workers Seasonal workers	1990 Decennial Census (available on CD-ROM)	U.S. Department of Commerce, Bureau of the Census, Suitland, MD 20233

Source: *Force, Jo Ellen, David Fosdeck, and Gary E. Machlis, "Monitoring Social Indicators for Ecosystem Management: The Technical Assessment Data for Oregon and Washington," Interior Columbia Basin Ecosystem Management Project under Order #43-0E00-5-5269, 1995, pp. 86–89.*

E·I·G·H·T

Using the Model for Science during Crisis

1

The application of the HEM has recently been extended to include science-based assessments during major environmental crises.[1] The model has been used as a scenario-building tool during both the Deepwater Horizon oil spill (2010) in the Gulf of Mexico and Hurricane Sandy along the East Coast of the United States (2012). These extraordinary events—one originating in failures within the social system (government, technology, and industry, for example) and one originating in a historical natural event—offer robust tests of how the HEM can be used to improve decision-making, resource management, and policy.

The Deepwater Horizon oil spill (the technical name is the MS252 oil spill) began on April 20, 2010. The Deepwater Horizon oil rig exploded, killing eleven oil workers and sinking into five thousand feet of water in the Gulf of Mexico. Oil continued to spill through April and early May.

On May 19, 2010, the Department of the Interior (DOI) established a Strategic Sciences Working Group with the objective of rapidly providing DOI leadership with science-based analyses of how the MS252 oil spill might impact the ecology, economy, and people of the Gulf of Mexico. The Working Group was not established to conduct a new or lengthy scientific investigation but to provide a rapid, science-based assessment of the potential consequences of the spill that could provide usable knowledge to decision-makers. The Working Group had several tasks: to quickly gather the existing relevant scientific information; to use this information and expert scientific opinion to develop alternative scenarios concerning the cascading consequences of the MS252 oil spill during the emergency response, mid-term, and long-term recovery/ restoration period; to share the results of this work with DOI leadership; and to test the usefulness of such strategic science working groups for other major environmental events. The Working Group (approximately fifteen scientists from different disciplines and organizations) convened its first session on May 25–28, 2010, in Mobile, Alabama, began delivery of scenarios to decision-makers within three days, and published its first progress report in August 2010, along with an overview paper in *Science*.[2]

The MS252 oil spill had potentially significant consequences for the ecological, economic, and social systems of the Gulf of Mexico (GOM). The Working Group treated the region of potential impact as *a coupled natural-human system* and approached the task of scenario building from this interdisciplinary view.[3] Hence, the Working Group did not limit the scenarios to separate biological, economic, or social consequences but included the way these consequences interact in shaping possible trajectories of the overall system.

Many alterative conceptual models of coupled natural-human systems exist in the literature. For the purposes of the Strategic Sciences Working Group, the HEM (as it appeared in 2010, see figure 8.1) was used in the scenario-building sessions. Its attributes were compelling: Reasonably detailed, the HEM includes both biophysical and socioeconomic elements, is explicit regarding flows, and has an emerging record of application, as we have shown.

The Working Group expanded the list of biophysical variables to cover conditions specific to the MS252 oil spill. This list (which includes several overlapping categories) is shown in table 8.1. The Working Group used the HEM as a checklist of possible relationships to ensure consideration of key elements of the coupled natural-human ecosystem for inclusion in the scenarios.

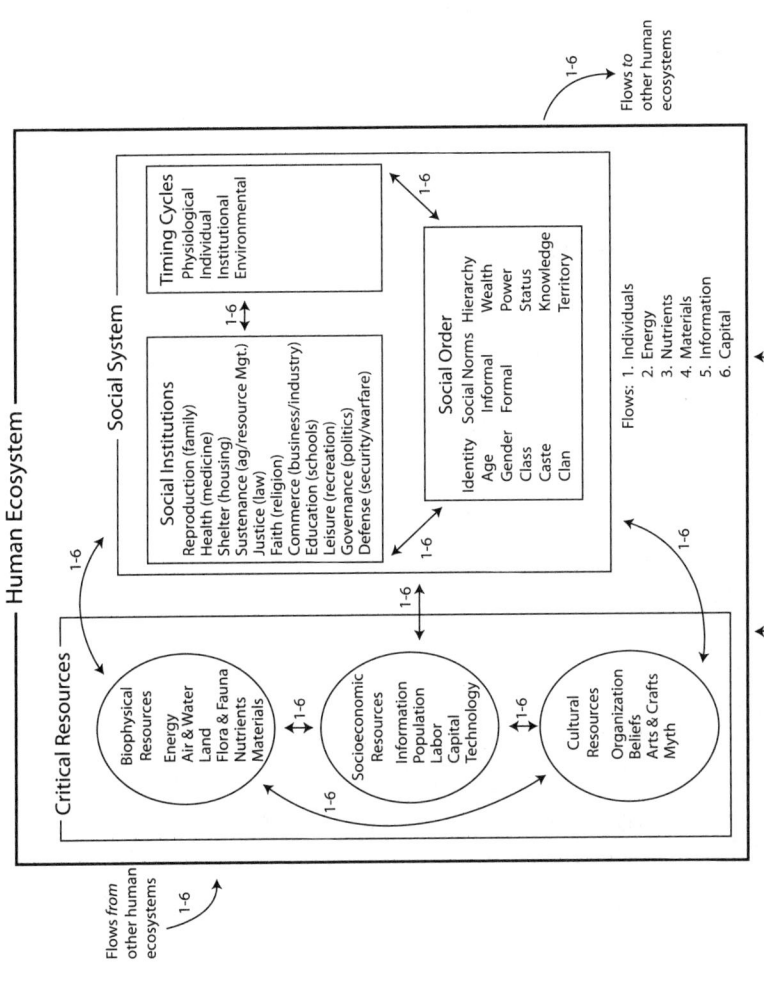

Figure 8.1. The Human Ecosystem Model as used by the Strategic Sciences Working Group for the Deepwater Horizon oil spill (from the Mississippi Canyon 252/Deepwater Horizon Oil Spill Progress Report 2, 2012)

Table 8.1. Selected additions to the human ecosystem conceptual model, biophysical resources

Flora/Fauna	Energy	Land
Plankton	Wind	Wetlands
Nekton (all kinds)	Solar	Uplands
Megafauna	Tidal	Beaches
Picoplankton	Electricity/natural gas	Barrier islands
Birds	Current	
Fish	Wave energy	
Submerged aquatic vegetation	Water	Materials
Marine mammals	Fresh water	Wood
Turtles	Salt water	Soil
Coral	Surface	Rock
Terrestrial wildlife	Salinity	Metal
Terrestrial animals	Temperature	Calcium carbonate
Domesticated animals	Depth	Plastic
Insects	Turbidity	
Forests		
Mangroves		
Grass beds		

From the HEM, the Working Group moved to develop a scenario framework adapted from the scientific literature on natural hazard response.[4] The scenario framework includes a general trend line of coupled natural-human system (the human ecosystem of HEM) stress over time divided into several key time horizons. The scenario framework is an idealized, conceptual one; other potential trajectories exist. The framework is shown in figure 8.2.

Within this scenario framework, the Working Group identified increasing baseline (pre-event) stress in the GOM prior to the MS252 oil spill. This reflects numerous known trends in stress: nutrient loading, expansion of the seasonal hypoxic area ("dead zone"), wetland loss

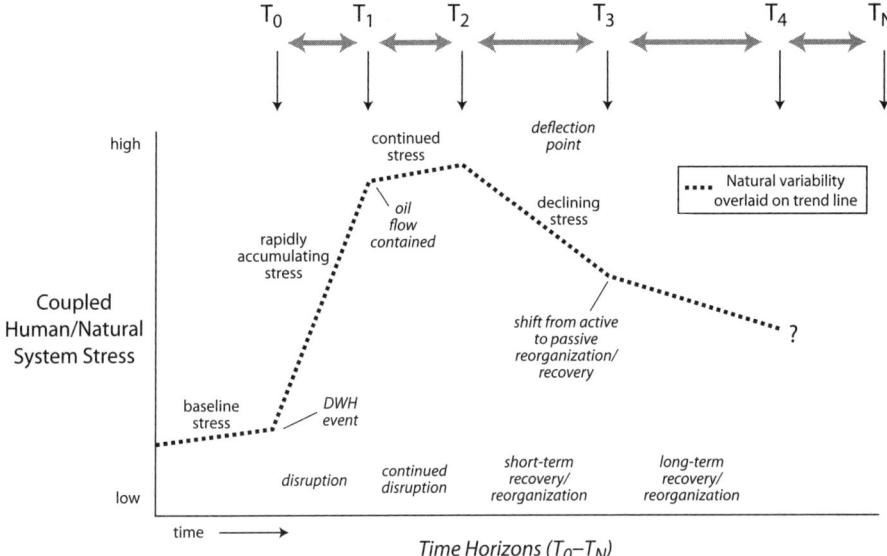

Figure 8.2. Scenario conceptual framework used by the Strategic Sciences Working Group for the Deepwater Horizon oil spill (from the Mississippi Canyon 252/Deepwater Horizon Oil Spill Progress Report 2, 2012)

and land subsidence, invasive floral and faunal species, climate change, increased fishing pressures, continuing effects of major hurricane damage in previous years, national and regional economic recession, and other factors.[5] These baseline stresses emerged from using the HEM Model as the checklist mentioned earlier.

At the time of the MS252 spill (April 20, 2010/T_0, identified in figure 8.2 as the "DWH event"), system stress was hypothesized to accumulate rapidly, initiating a period of significant system disruption. After the oil flow was contained (July 15, 2010/T_1), system stress was hypothesized to continue to rise due to a series of lagged effects such as landfall of previously released oil, re-release of sequestered oil and dispersant, or chronic toxicity to sensitive ecosystem components. At some time in the future (T_2), system stress is hypothesized to decline (the "deflection point") due to a combination of reduced inputs of stressors; natural and social resilience in the coupled natural-human system; active emergency and recovery responses by national, state, and local entities; and other factors.

Further along the time horizon (T_3), the stress trend further deflects, as short-term recovery/reorganization (with its active and adaptive responses) gives way to long-term recovery and passive response. Examples of passive responses might include water quality improvements or economic redevelopment without substantial government or industry intervention. T_4 and T_N represent longer-term time horizons over which recovery processes may persist. These time horizons are derived from the HEM's treatment of timing cycles, and may vary significantly in duration measured in days, months, years, or decades.

Within this HEM-based framework, there is an assumption that recovery often involves some reorganization of the system rather than a full return to a preexisting state. Baseline stress in these future horizons is largely unknown. Figure 8.2 illustrates that natural variability (both spatial and temporal) is overlaid on general stress trends, and thus care should be taken to distinguish between responses to the MS252 event and natural variability or "noise" in a system.[6]

During its first session, the Working Group established several geographic or spatial units to help focus the alternative scenarios, incorporate distinctive consequences associated with the different units, and provide useful information to decision-makers. These units were also used in the second session and included vertical life zones, major ecosystem types, sociopolitical and administrative units from the local village to parish, county, state, and biodiversity quadrants (here the Gulf of Mexico).[7] Again, these reflect the interdisciplinary nature of the HEM.

The Working Group established a common method of scenario building. It began with training the Working Group on the background, context, and use of the HEM. Next, an initial condition resulting from the selected scenario parameters was established, such as "persistence of oil and dispersant in months in the mid-water life zones in the NE biodiversity quadrant." From the initial condition, the group developed a set of cascading consequences, first using the HEM to identify key variables and cascades, then through the sharing of expert opinion, scientific literature, and in-depth discussion. These cascades were informally drawn on whiteboards and simultaneously entered into a graphic program called SmartDraw.[8] SmartDraw enabled the Working Group quickly to modify and expand on existing cascades as well as periodically revisit the HEM.

A key element of the Working Group's task was to assign preliminary levels of scientific uncertainty to each of the cascading consequences. These reflect the state of knowledge for complex and significant disruptions in human ecosystems (which can vary from substantial scientific certainty to unstudied and unknown relationships), the state of knowledge for the specific system (GOM) and its system functions and processes, and the need to provide decision-makers with a practical method of assessing levels of uncertainty for policy and decision-making.

Following Charles Weiss, several alternative scales were considered: legal standards of proof, informal scientific levels of uncertainty, Bayesian probabilities, and the climate change–specific scale adopted by the Intergovernmental Panel on Climate Change (IPCC).[9] The Working Group adapted the Weiss scale of informal scientific uncertainty, as it is well suited to scenario building and allows for systematic refinement as new information becomes available (a key characteristic of the MS252 event). In the Working Group adaptation, several of the Weiss scale categories were aggregated for clarity and to allow for rapid assessment. Table 8.2 illustrates Weiss's original scale and the Working Group's adaptation. Following the development of a specific scenario, the Working Group established uncertainty levels (0–5 and not known) for each cascading consequence within the scenario. Individual Working Group members with appropriate expertise provided opinion bolstered by review of the available literature and contacts with additional subject matter experts.

The scenarios provided decision-makers and resource managers a set of possible intervention points at which they could focus attention on key interventions likely to have substantive effect on reducing negative impacts (such as re-release of sequestered oil). Interventions may also offer the ability to increase resilience and positive recovery

Table 8.2. Levels of scientific uncertainty

DOI Strategic Sciences Working Group Categories	Weiss (2003) Informal Scientific Categories
5—certain	Certain
4—reasonably certain	Very probable + reasonably certain
3—probable	Likely + probable
2—plausible	Possible + probable (more info needed for firm conclusions)
1—unlikely	Unlikely (supported, but not entirely ruled out)
0—not possible	Not possible (violates established laws, theories, principles)
nk—not known	Insufficient information to ascribe level of certainty

responses (such as improved monitoring and targeted income support). This is particularly useful during the long-term recovery period and could help accelerate sustainable recovery. DOI leadership requested that the Working Group identify possible interventions. The HEM was also used to help identify candidate interventions.

The scenario-building technique, based on the HEM and employed by the Working Group, has of course several limitations. The scenarios are not quantitative risk analyses or predictive models. This approach does not include detailed linkages and feedback loops among different components and does not use environmental endpoints and values. All possible trajectories cannot be anticipated. The assigned scientific uncertainties reflect the scientific literature and expert opinion for each individual consequence in a chain; summary uncertainties for a full chain of consequences or a scenario were not assigned. The scenarios are not spatially specific at scales other than the identified GOM quadrants and spatial units used as a parameter. Like most scenario building, the scenarios are constrained by available expert opinion, information, and theory.[10]

Yet the scenarios provided decision-makers with specific and valuable insights. Scenario 4 of the Working Group examined the time period from the mid-term to long-term recovery/reorganization, during which it is expected that stress in the system would be declining. Scenario assumptions were: 3.2 million barrels of oil remaining in the Gulf System and at least one major landfall tropical storm or hurricane during recovery. The scenario parameters chosen by the Working Group were: toxicity of oil and dispersant persisting for years in the northern biodiversity quadrants of the GOM, coastal communities as the spatial unit, and T_2-T_4 as the time horizon. The scenario is shown in figure 8.2.

Numerous potential direct consequences were identified, including: contaminated Gulf seafood, continuing human exposure to oil and dispersant, contamination of coastal wetlands, fish mortality, contamination of the benthic life zone, contamination of the pelagic life zone, depletions of marine/estuarine populations, behavioral response by animals, continued contamination of beaches, stressed wetland flora, continued contamination of barrier islands, and other post-spill activities.

A strategic science response to a disruptive event like the Deepwater Horizon oil spill can provide immediate assessment of the range of system stresses and the priorities for effective restoration and reconstruction. This enables a nearly concurrent restoration response, which can lower the peak stress and also accelerate the overtime reduction in stress, which can prove particularly beneficial in the event of a secondary disruptive event.

In addition, the technique is well suited to provide scientific assistance to preparations, emergency response, and recovery efforts related to other emergency incidents, including large-scale oil spills, bioterrorism attacks, hurricanes, earthquakes, significant wildfires, floods, and other hazard events. There may be a unique and valuable role for this form of strategic science. In January of 2011, Secretary of the Interior Ken Salazar issued Secretarial Order 3188, formally establishing the Department of the Interior Strategic Sciences Group. And at its core is the HEM.

2

In late October 2012, Hurricane Sandy advanced toward the eastern seaboard of the United States as one of the most severe storms to ever threaten the region. At the time of its U.S. landfall near Atlantic City, New Jersey, on October 29, Hurricane Sandy measured over 1,100 miles in wind-field diameter and was classified as a post-tropical storm with maximum sustained winds of 70 miles per hour. Fueled by a coincident nor'easter and spring tides, Hurricane Sandy directly affected seventeen states with storm surges of up to 8.57 feet, heavy snowfall (over three feet in parts of West Virginia and North Carolina), and historic flooding. The storm was directly responsible for at least seventy-two deaths in the region, and thousands of individuals were displaced from damaged or destroyed dwellings. Over 8.5 million households lost power, and the United States sustained approximately $50–$70 billion in damages, making Hurricane Sandy one of the costliest storms to ever strike the United States.[11]

In response to the effects of Hurricane Sandy, on December 7, 2012, President Barack Obama established the cabinet-level Hurricane Sandy Rebuilding Task Force. In January 2013, Secretary of the Interior Salazar directed the Strategic Science Group (SSG) to deploy a crisis science team to support the Department's role on the Task Force. In response, the SSG assembled a team of experts (again, about fifteen scientists) from government, academic institutions, and nongovernmental organizations—Operational Group Sandy (OGS)—to develop scenarios for the Task Force to examine the short- and long-term impacts of Hurricane Sandy and future major storms (such as another major hurricane) on the ecology, economy, and people of the affected region. Again the HEM was used as a foundation for the operational team's work.

The OGS defined two "regional types" for their scenarios to account for the geographic heterogeneity in the affected region. Coastal Communities and Ecosystems were areas that were completely or partially inundated during Hurricane Sandy and/or within 0.25 miles of the coast (for instance, Fire Island, New York). Urban Communities and Ecosystems were distinguished from Coastal Communities by high population density (>50,000 people/square mile), high building density, high average building height, interconnected and underground infrastructure, "armored" shoreline (such as breakwaters and riprap), and high socioeconomic diversity (for instance, Manhattan). There is some overlap between these two regional types.

As with the Deepwater Horizon Oil Spill, Operation Group Sandy used the HEM to focus the scope of its scenarios. For the Hurricane Sandy scenarios, stress to the human ecosystem was defined as heightened consequences, such as thirst in the face of lack of water, and/or increased requirement for adaptive responses, such as water conservation in the face of drought. Both heightened consequences and the increased requirement for adaptive responses are cumulative over time and can spread through multiple variables and flows within the human ecosystem, hence the utility of the HEM as conceptual model and organizing framework.

To build the Hurricane Sandy scenarios, the OGS followed methods developed by the earlier Strategic Sciences Working Group during the Deepwater Horizon oil spill.[12] These methods included developing detailed scenarios to illustrate important cascading consequences, assigning a qualitative level of uncertainty to each consequence, and identifying potential interventions that would improve the resilience of the coupled natural-human system to future natural hazards. Assigning a value to each intervention ("an intervention value") was a new technique introduced for the OGS session.

Figure 8.3 shows an annotated chain of consequences as a partial example of the SSG scenarios. In this chain, one of the consequences of Hurricane Sandy was evacuations. Evacuations resulted in the disruption of daily behavioral patterns, which led to mental and physical stress, heightened stress to local service providers, and ultimately to continued physical and mental health impacts.

After the scenario-building process was complete, the OGS developed interventions, defined as institutional actions that support the recovery as defined by the Task Force and that increase the resilience of the coupled natural-human system to future major storms. Each intervention was assigned an intervention value of High (H)/Medium (M)/Low (L), which is a subjective index of the potential return on investment (not only monetary) *and* pervasive impact on the system.

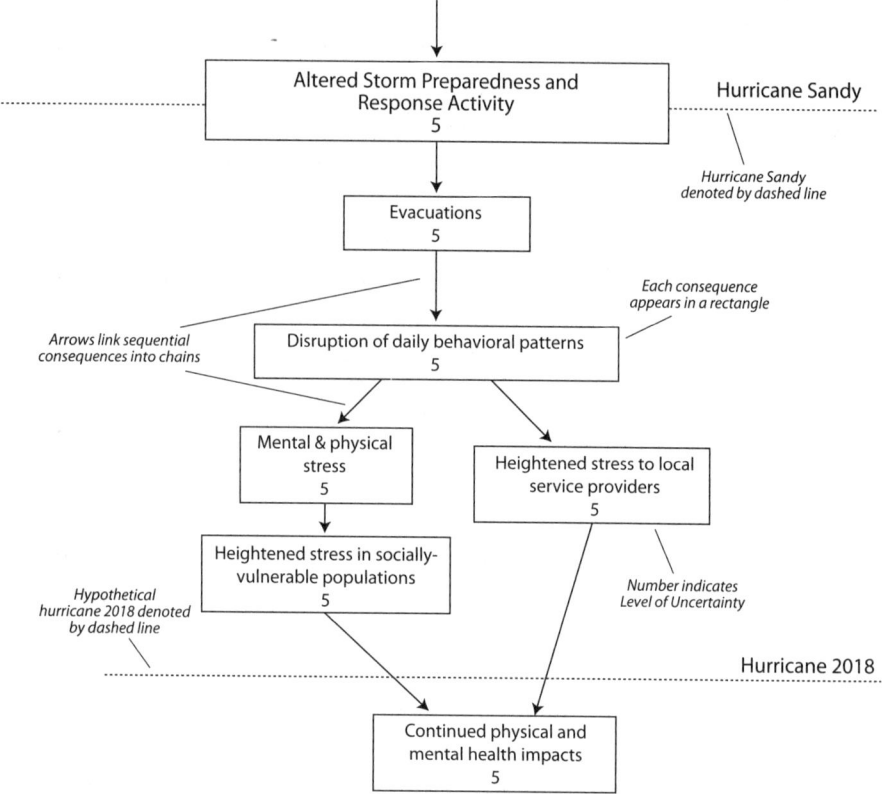

Figure 8.3. Schematic of SSG scenarios illustrating cascading consequences. Each consequence is assigned a level of scientific uncertainty. Hurricane Sandy and hypothetical Hurricane 2018 are illustrated using dash lines. (From the Operational Group Sandy Technical Progress Report, 2013)

Pervasiveness was defined as the number of first-tier consequences that are positively impacted by the intervention.

The OGS scenario of Hurricane Sandy's impacts on Coastal Communities and Ecosystems identified a total of thirteen "first-tier" consequences that were a direct result of Hurricane Sandy in these regions. These consequences span a range of environmental, economic, and social effects within the HEM, and include: ecological change; changes in coastal geomorphology; atypical fresh/saltwater mixing; flood damage to the built environment and property; wind damage to the built environment and property; loss of electricity; disruption of commercial and recreational fishing; closure of outdoor recreation resources; altered storm preparedness and response activity; injury, stress, and loss of human life; altered perception of risk; increased voluntary activity; and altered beliefs and values.

Use of the HEM provided insights on chains of consequences. Hurricane Sandy led to changes in storm preparedness and response activity, especially those that are flood-related. Cascading consequences include reasonably certain reduction and prevention of damages in the built environment, observed enhanced preservation of communication networks, restricted access, increased expenditures at multiple levels, and evacuations.

Increased expenditures for preparedness at the instructional, commercial, and household levels will certainly lead to the pre-positioning of emergency personnel, critical equipment and supplies, and storm supplies and food for future major storms. Consequences of evacuations included the disruption of daily behavioral patterns, which precipitated observed heightened levels of stress to local service providers and mental and physical stress to the affected population, with the reasonable certainty of more adverse effects on the socially vulnerable, such as the elderly, sick, young, or pregnant. The results suggest that in the long term, these consequences will probably lead to physical and mental health impacts, which has implications for the coupled natural-human system during future major storms.

Hurricane Sandy's impacts to the social dimensions of the coupled natural-human system also included observed injury and stress. This led to multiple cascading consequences including the observed increased demand for medical treatment, services and supplies, increased health care costs, increased use of social services, observed emotional trauma on multiple levels (from household to community and institution), and loss of household income.

The increased demand for medical treatment and supplies has multiple cascading consequences. These include the reasonably certain diminishing but ongoing demand for treatment, support, and services as the community recovers, and probable new demands for medical services as long-term health impacts of Hurricane Sandy emerge.

The SSG scenarios, based on the HEM as well as the scenario-building techniques employed during the Deepwater Horizon oil spill, provided several key insights:

- The consequences of Hurricane Sandy are complex;
- There are substantial uncertainties associated with both the consequences of the storm and potential interventions to improve resilience against future major storms, and storm response activities should account for this uncertainty;
- Resilience is best achieved when developed for the coupled natural-human system rather than by applying measures only to individual units or infrastructures;
- Both "gray" and "green" infrastructure are necessary for improving resilience; and
- The speed and effectiveness of interventions may have substantial impact on the capacity of the region to increase resilience to future major storms.

Upon completion of the scenarios, the SSG staff briefed Task Force representatives, DOI leadership, White House staff, and officials in the affected region. DOI has since used the findings to inform the prioritization of $300 million in supplemental mitigation investments to enhance regional resilience. On August 19, 2013, the Hurricane Sandy Rebuilding Task Force released its report to the President detailing a rebuilding strategy for the Hurricane Sandy–affected region. The report includes sixty-nine policy recommendations, some of which flowed from the SSG work based on the HEM.

Multiple federal, state, and local authorities continue (even now in 2016) to debate the best paths forward to rebuild the Hurricane Sandy–affected region in ways that enable the area to rebound quickly and that simultaneously bolster the region's resilience to future storms, rising sea levels, and other hazardous events. Proposed changes have included land buyouts and planned retreat, green and gray infrastructure fortifications and design competitions, and new social services and volunteer coordination networks. Hurricane Sandy helped reignite the national conversation on how the country should adapt to rapid urbanization and changes in global climate to better prepare and protect its citizens, infrastructure, and natural resources for the future.

The Deepwater Horizon oil spill and Hurricane Sandy events, and the deployment of the SSG in both cases, served as valuable tests of the HEM. They demonstrated that in addition to supporting resource management, urban studies, and policy-making, the HEM can support science during crisis.

N · I · N · E

Revitalizing Human Communities and Reclaiming Biological Communities
The Baltimore Story

1

The HEM provides a unity of understanding with shared concepts, a framework, and a model for resolving complex human ecosystem problems. With it, decision-makers from different organizations—public and private—may coordinate their work with that of local citizens. The emphasis is on the whole system, which combines issues such as trends in crime, housing, education, health, natural resources, and community stability into an integrated network. This chapter illustrates how the framework and model was applied in a major city in the United States.

Our learning was based on working with and learning from professionals and local people residing within the 276 neighborhoods of Baltimore, Maryland—an older, mid-size, northeastern U.S. city that is in transition from an industrial to an information political economy. When the Urban Resources Initiative (URI) started in the city in 1989, there was an infrastructure to serve a million and a half people, but the city's actual residential population was only around 740,000.[1] Abandoned buildings and depressed neighborhoods were the norm. Heroin epidemics were common, and crack cocaine was rising as the drug of choice. Unemployment among black males was high, and an available occupation in many neighborhoods was dealing drugs. Urban "renewal" developments destroyed traditional neighborhoods with their low-cost housing and "eyes on the street" social control, substituting housing projects that were often incubators of crime and a form of incarceration for the elderly poor.[2] Most of that housing has been torn down and replaced with low-rise housing. The poverty rate has remained the same: 22 percent in 1990 and 23 percent in 2014.

Though having many of the problems of any city as it changes from an industrial to an information economy, Baltimore has many assets. It has a revitalized harbor area. It has great universities like Johns Hopkins and Loyola. It has an Olmsted-inspired park system of nearly six thousand acres. It has a substantial number of white and black middle-class citizens determined to renew the city's well-being. It has impressive museums, churches, and historic buildings. It has a notably high number of block watch associations. It has two major professional sports teams—the Ravens football team and the Orioles baseball team. It has Pimlico Race Course, where the Preakness, the second leg in the Triple Crown of thoroughbred racing, is run. It has the Parks and People Foundation—a community-serving organization that is an icon for the value of voluntary nongovernment organizations serving the city. The National Park Service manages the Fort McHenry National Monument and National Historic Shrine. This monument was significant in U.S. history, as it was where the Americans outlasted the British in September 1814. And it is where Francis Scott Key wrote the poem that later became the national anthem. So there is a historic reminder of recovery and renewal as part of the culture of the city.

In 1950, the city's population was 949,708 and had grown by 10.5 percent since 1940. From that peak the city has lost people every decade, dropping to 620,961 in the 2010 U.S. Census. The 2013 estimate indicates that the city had a slight increase in population to 622,104, that is, 1,143 residents more than in 2010. The percentage of the population that is African-American has increased, and the white population has declined. Today 63 percent of the population is black, 30 percent is white, and 4 percent is Hispanic or Latino. As the suburban population ages and new housing along the waterfront expands, there is likely to be a return of white residents in the city center.

The Baltimore story emphasizes that certain universal problems and solutions confront all human societies. Those bright and shining new cities in China with populations in the millions will soon reach their "middle age," when maintaining rather than growing is the dominant challenge. Further, adaptation is not simply a planning abstraction but involves significant transformation of the "trained incapacities" often found in the cultures of most public and private agencies. Thus, part of the task of the URI team was helping the recreation and park agency move from a custodial to a more activist stance and become the center for community development and environmental restoration for the city. The universality of problems and the search for integrated solutions required a framework like the HEM to identify, apply, and store learning. The goal was to find and propose ideas and practices that ensure a more sustainable, efficient, and effective civic order in Baltimore and elsewhere. Still, as one travels the world of conferences, meetings, and symposia on the new urban century, there remains a faith that some magic architectural bullet or some super-technology will emerge to guide planners and managers of these urban ecosystems. The most popular seems to be the geographic information system (GIS), which is an excellent guide to gross clustering and trends in land use and social life but often fails to connect this grand-scale technology with actual information about behavior and hopes.

One early lesson was that there are no easy technological fixes. Though tools like GIS do a good job of mapping spatial patterns and relationships, they are mostly descriptive rather than problem-solving. Thus, these data are useful abstractions of the geography of a place but do not fully connect to the reality in which most people live out their lives. There is the need to move from planning documents to the ground level where emotion, motives, rituals, and constraints actually drive the regularities of human ecosystem behavior. It is often the case that a simple hand-drawn map of a district or neighborhood is more helpful in managing crime, responding to environmental crises, or issues of human health. Indeed, it may be of more analytic value than the high-tech, high-elevation map. The more humble map puts less distance between the central authorities and their clients and makes the knowledge of local people a more respected, robust, and effective part of the decision process. The conferring of legitimacy to local knowledge is a critical factor in sustaining cooperative efforts when the central authority is disconnected from communication linkages with local people, such as during an emergency or environmental crisis like floods, fires, hurricanes, epidemics, wars, and violent weather events. Large-scale maps are very useful for identifying most at-risk zones but are limited when one needs to know how to help people get out of the flood about to wash over their house. That kind of organization is always local.

As noted above, the direction of work in Baltimore was organized by the HEM; its application was shaped by the empirical wisdom given by colleagues in the Department of Recreation and Parks, the Parks and People Foundation, the insights of interns, and the regular people in the neighborhoods themselves. Much insight was gained from colleagues in other state, municipal, and federal agencies, and scientific partners in the National Science Foundation—Long Term Ecological Research (LTER)—Baltimore Ecosystem Study (BES). Some of the URI team's efforts were based on social action research, some were trial-and-error decisions, though most were shaped by the discipline and structure of science. Many of the lessons learned in this effort should help researchers, policy-makers, planners, and managers in other urban regions around the world apply the ecosystem approach and the HEM in managing their cities.

2

Understanding the lessons learned from the Baltimore Urban Resources Initiative for co-development of communities and ecosystem restoration requires some background on the program's origins. In 1989, Burch and Dr. Ralph Jones served on a national committee to explore the future needs and directions for research and policy in the U.S. National Park Service. Burch had just assumed management of a large development project in Nepal, and Dr. Jones had just been named director of the Baltimore Department of Recreation and Parks. Burch's work was centered on community-based resource work in South and Southeast Asia, and the mission in Nepal was to infuse this approach into the central natural resource training institution. Dr. Jones was confronted with a greatly reduced organizational capacity facing increasing demands for services and a passion for having his department be a force for community stability and well-being. Therefore, they found a great deal of similarities between the poor urban environments and the poor rural environments of Asia, where Burch usually worked. In both cases, a poor environment and poor people fed on one another, leading to mutual degradation. The trick was finding the means to break that cycle. Burch discussed some of the participatory and empowering strategies he was using in rural areas of Asia and other developing regions. Dr. Jones insisted that these strategies might work in his city and challenged Burch to take up this issue. As details were being worked out, Dr. Jones tragically had a heart attack and died. The Initiative pushed ahead under the new and more skeptical department leadership to fulfill the URI promise to Dr. Jones.

This was a participatory cooperative effort between Baltimore's Department of Recreation and Parks and the Yale School of Forestry and Environmental Studies. Dean John Gordon provided fiscal and intellectual support to the collaboration, and the Department of Recreation and Parks was encouraged by the office of Baltimore mayor Kurt Schmoke to develop the connection. Three politically savvy women with activist ties to people in the neighborhoods—Sally Michel, Alma Bell, and Laura Perry—thought the collaboration had great promise. It was a mutual learning effort that sought to have the Department of Recreation and Parks become a community development and serving organization through rehabilitation of the natural and social environments. At the same time graduate students at Yale gained a substantial education in how to work with and within human ecosystems. The core mantra was that biological ecosystem rehabilitation can only occur and be sustained by the level of community revitalization.

The mechanism was a reconfiguration of the agency and the role of parks and vacant lots; more importantly, it fostered the notion that a specific tree or patch of grass is not just a bit of street furniture but rather a place for mutual learning that could aid the organization and create the ability and pride of local people to take charge of their environments. In a developing rural area, a tree might be seen as a provider of fuelwood or fodder or sawtimber; in the city, it was the tree's symbolic and social value that was the highest and best use. Planting and looking after a tree or a garden announced that the neighborhood had pride of place, and the act of having the tree brought people together around a shared resource.[3] In Baltimore and other urban areas, a tree, woodland, or forest isn't merely a product; it is more a means for strengthening the local community. In Nepal and other rural areas, the tree is primarily an end for producing timber, fodder, and fuelwood.

The URI team's action research required attention to the shaping of the agency and identifying the most likely ways to explore connections between greening and community revitalization and the larger context of these activities. The first task was to identify, from the existing 276 Baltimore neighborhoods, the nature, types, trends, characteristics, assets, and constraints affecting the twenty-five or so neighborhoods in which they were to work. Further, all participants had an interest in diminishing nonpoint source pollutants entering the Chesapeake Bay, which served as the base biological indicator. Baltimore citizens cared because so much of the municipal identity and way of life of Baltimore was tied to the health of this great ecosystem.

Following the Hubbard Brook studies guidelines, the team needed to identify ecological units that broke out of the grid and joined neighborhoods to a common resource and whose boundaries were more functional than the usual political and census units. In Baltimore the URI team identified neighborhoods associated with the three municipal watershed units—Gwynns Falls, Jones Falls, and Herring Run—and documented how these neighborhoods interact with the hydrological flows, soils, and botanical patches associated with the watersheds. Emergent questions concerned how household and neighborhood "ecoprints" vary and how variation in energy, nutrient, material, and hydrological flows related to the observed social structural patterns directing these flows. The URI had need for optimistic and enthusiastic workers; it turned to student interns, willing neighborhood volunteers, and idealistic members of the agency and the city at large.

Clearly, this was a complex system. To avoid running in all directions at once a way to organize was necessary, to separate the critical from the less critical connections, and to focus work so that there was some steady improvement in the life of the city. For this the URI team turned to the HEM to identify the range of variables and their scale, relationships, and key points of intervention. Of course, there are many available models and frameworks. For example, Rusong Wang and Zhiyun Ouyang have provided a large-scale, complex framework for planning in Tianjin, China. D. J. Rapport and A. Singh have a straightforward grouping of broad, abstract factors driving the human ecosystem.[4] These are but two of the many models available, each of which has a particular utility. However, most of these models do not permit the analyst to work at the local household and community level of behavior. The HEM encourages the connection between the larger policy abstractions and the reactions at the local level. This application of HEM is primarily a planning and management effort and, though work in neighborhoods was critical, the plan was broadly at the city-wide mosaic level.

Figure 9.1 indicates the set of variables and their clusters, linkages, and flows; it highlights the variables considered critical in planning the Baltimore action research effort. The following discussion describes the actions tried out and the specific variables under which they fit. Remember that all the variables are important and need to be considered. However, limits of time and reality suggested that some variables were more essential than others. We hope that the reader will move between the figure and the text so that the conceptual mapping will be clearer.

3

The URI team first looked at the critical resources for the Baltimore study. Four variables in the biophysical component were emphasized—*water, land, flora,* and *fauna.* The team's concept of water, its quality, volume, and basic hydrology, reflected the ecosystem research design followed by Bormann, Likens, and their associates, and Golley.[5] Baltimore grew up around three watersheds—Herring Run, Jones Falls, and Gwynns Falls—and these definable boundaries shaped the kinds of development, settlement, and transportation issues affecting the human ecosystem sub-units. An important problem was the influence of waterborne nonpoint source pollution entering the Chesapeake Bay from the city. Watersheds, of course, influence and are influenced by the adjacent

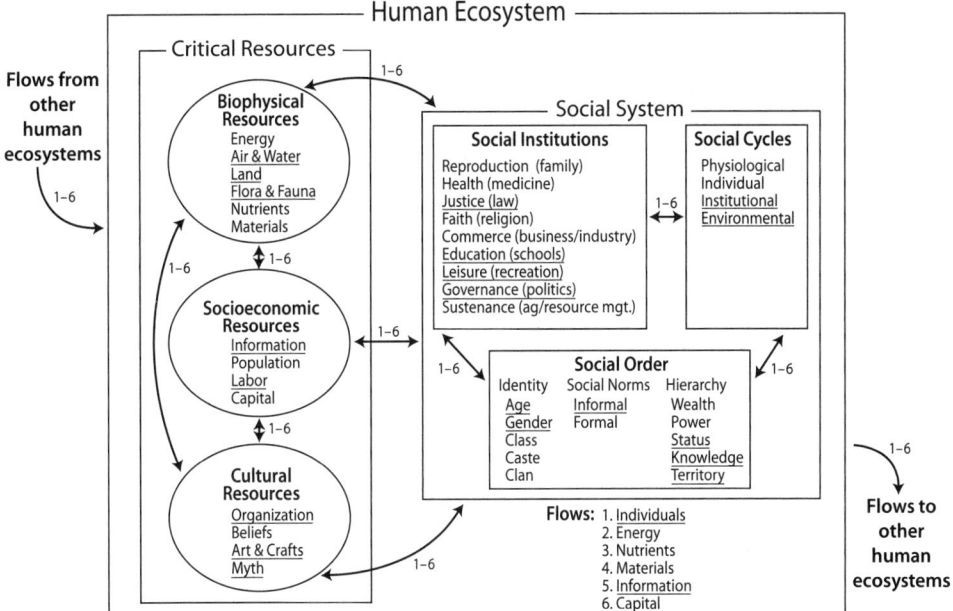

Figure 9.1. Rehabilitation and revitalization of community ecosystems: some key HEM variables for guiding action in Baltimore, Maryland

neighborhood land and the land uses of factory, housing, streets, golf courses, parks, and so forth.

A watershed example described in figure 9.2 illustrates the multivariate nature of most human ecosystem issues. The first column illustrates the zone of awareness that socially defines an environmental problem that often emerges when citizens and persons in powerful political positions use an aesthetic indicator to stimulate and support their awareness and concern. The usual response is a quick biophysical or engineering solution, and that is where it usually ends. This is the zone of analysis and action where dependent and intervening variables direct the remedial action. Figure 9.2 then moves the "cause" to considering second- and third-order "causes" that must be addressed if the solution is to be sustainable. However, as noted, this is usually the zone of inattention if the engineering quick fix seems to hold in the short run. Yet, it is these underlying "causes" identified by the HEM categories such

Zone of Awareness	Zone of Analysis and Action (Critical Resources)		Zone of Inattention (Social)	
Concern Indicator (Biosocial)1	Dependent Variable (Biophysical)	Intervening Variables (Socio/cultural)	Independent Variables (Sociaol Norms)	Facilitating Variables (Public Policy Social Cycles)

Fish Dying (Fauna)
 	
Water "looks" dirty (Water)
 	
Phone calls to DEP (Information)
 	
Mayor tells R&P to fix it (Organization)
	Rapid and above nromal sediment loading in Gfalls- measured at 3 gauging stations. (Information cycle)
	
	(Land-Water Cycles) Sediment Load:
		Substantial perturbation in upstream watershed vegetative cover- related to construction of 120 unit residential development on one- acre lots in a 210 acre tract. (Organization)
		
		Development scale:
		  Small = 1–11 units per acre
		  Modest = 12–22
		  Large = 23–33
		  V. Large= 34 and up
		Land-Status
		
		Terrain scale:
		  Slope 1–5
		  Soil 1–5
		  Vege 1–5
		
		(Land)
			Zoning variances eased; (Justice Cycles)
			
			Interest rate changes; (Capital Cycles)
			
			Banks need to circulate surplus (capital);
			
			High levels of employment; (Labor)
			
			Family life cycle demographics favorable; (Population)
			
			County development; (Governance)
			
			Agency markets site; (Commerce)
			
			Developer has project in area-and can pyramid labor and equipment better when several projects at different stages in development cycle are close at hand. (Labor-Capital Cycles)
				Highway subsidy (Capital)
				
				Morgage deduction (Capital)
				
				County subsidy (Governance)

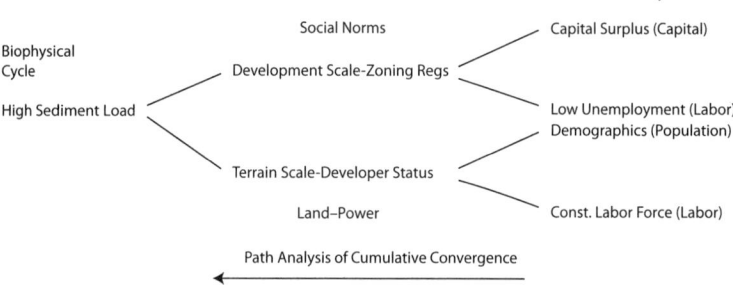

Figure 9.2. HEM analysis of second- and third-order causes of specific ecosystem events

as *justice* and *governance* social institutions that have significant consequences when they intersect with critical resources such as *information* and *land*. Here an ecosystem approach with strong emphasis on human social variables is necessary for developing truly sustainable policy and management solutions.

The diagram at the bottom of the figure 9.2 gives an example of decision paths that affect the observed outcomes. When the URI first began work in Baltimore, the county had one of the highest deforestation rates in the world, as real estate developers expanded residential and strip commercial sites. There was a congruence of factors that led to this development; figure 9.2 permits considering the downstream effects in Baltimore's three watersheds, where thirty-year floods were occurring in short time scales. The path analysis permits some structured prediction as to likely outcomes resulting from decisions made to serve single-purpose outcomes, such as making mortgages available to first-time homeowners. These decisions passed on very high costs to the watersheds of the city.

Since the Department of Recreation and Parks owned a considerable portion of the stream banks along the watersheds, it was a natural choice and place of entry for a new proactive stance. There were vacant lots, abandoned structures, brown fields, mini-park properties, and the edges of streams that were being mowed and had few trees. Here was a chance to work on rehabilitating the landscape and revitalizing the neighborhoods as a mutual, reinforcing activity. *Flora* and *fauna* became important. "No-mow" meadowlands were established on the slopes going down to the streams. Trees and shrubs were planted here and up along the streets and in restored vacant lots. The removal of tons of accumulated garbage in vacant lots and the conversion of these lots to mini-parks and vegetable gardens helped to contain volumes of rainfall and reduced nonpoint source pollution. Further, with the change from vacant lots to parks and gardens the rats and other vermin had lost their food source and moved on to other sites; chipmunks and other more desired animals settled into the parks. Feral animal control was enhanced through reduction of den and hiding habitats, and the formerly derelict land uses now began to look appreciated, green, and managed, reflecting that circumstances had improved.

These biophysical critical resources were matched by two critical *socioeconomic resources—information* and *labor*—and three *cultural resources—organization, arts,* and *myth*. Interns from Yale and colleges and universities in the Baltimore area met an affordable labor need, but more importantly, they created new knowledge about many aspects of parkland ecology within the city. They did research on issues like management of climbing vines, the management of waste leaves, the best practices for communicating ecological understanding to local schoolchildren and neighborhood residents. Geographic information systems were introduced to the Department of Recreation and Parks, and maps of key parks and the park system at large were produced for distribution to the public. This information helped emphasize the wonders of nature right in their back yards. The URI, with co-parentage of Yale, the Department of Recreation and Parks of Baltimore, and the Parks and People Foundation emerged as the human ecosystem research and development wing of the City's Department of Recreation and Parks, coming up with innovative ideas, trying them out, and then reporting on their burdens and benefits. Interns became teachers and mentors for outdoor education and employment activities done jointly with another private organization—Outward Bound. This morphed into a program called Kids Grow that worked with children in poor areas after school using the parks as the training base. Dr. Shawn Dalton, a URI intern at the time, led in the development of the outreach program. The hope and realization was that these "kids" would connect their work on improving the neighborhood environment to protecting the Chesapeake Bay. For example, little blue signs were painted on storm sewer outlets in the neighborhood that read, "Don't Dump, Save the Bay."

Interns and Department professionals developed the first-ever master plan for the Department of Recreation and Parks.[6] To carry out the mandate of the plan, interns, again led by Dr. Dalton and professionals at the Department of Recreation and Parks, developed a training program for all employees that stressed the theme that all residents were stewards of the Chesapeake Bay and the Baltimore streams flowing into it. A training manual was developed to go with the plan and program as well as a series of short guides to the importance, characteristics, and uses of watersheds in the city. Trainees saw an empirical system as they followed the water flow from city-owned reservoirs in the county to treatment plant to use in the city neighborhoods to sewage treatment plant and out into the bay. Other labor actions were mobilization of interns and underemployed youths to work on vacant land clearances with local "block watch" groups and the conversion of these lots to usable spaces.

This meant mobilizing volunteer labor in the neighborhood as the prime beneficiaries of the gardens and their produce. Volunteer tree steward groups were organized and trained by the city arborist and interns to look after street trees and to plant where trees were missing. In short, knowledge, training, and action were preparing people on the ground to connect to the critical resources of land, streams, plants, animals, neighborhoods, and the Chesapeake Bay. This mosaic of opportunities internal to the Department of Recreation and Parks and those infused from reaching out to broader constituencies in poor neighborhoods, university students, and urban redevelopment groups provided mechanisms and context for restructuring the organization of the agency. A variety of publications and working papers track these activities.[7]

Along with organization, two other cultural resources were deemed most critical—the arts and myths used to guide public and private groups. The struggle to create the Gwynns Falls Greenway Trail and to establish opportunities for service and education for youth, elderly, and other groups in the neighborhood revitalization required the vision and the ability to re-envision that greening was not some aesthetic conceit imposed by the past on the present. Rather, green activities were visualized as a robust means for taking back neighborhoods, inspiring young people, reducing crime, and enabling dialogue between the generations and ethnic groups. A central force in this was the Urban Arts Institute of Baltimore, particularly the dedicated work of Steffi Graham, a photographic artist. She captured the ideals of the project, the people working in the neighborhoods, and monitored the progress and losses of the effort. Today there are still regular arts programs to bring people onto the greenway and to engender support for its maintenance and expansion of services. The work of painters, photographers, dancers, sculptors, actors, and landscape architects have all been enlisted in these greening activities in Baltimore rehabilitation and many similar areas in Europe and Britain.

This visual re-presentation has been part of the means for changing the *myths* about the Department of Recreation and Parks held by employees and outsiders alike—that its history was not one of new initiatives but a custodial organization that kept the place tidy and looked after the timeless Olmsted legacy. As noted earlier, the goal of Dr. Jones was to shift this comfortable mythology that infects many urban parks departments, which assumes that open space and recreation services should sit and wait for their clients to appear, to a proactive, creative, and dynamic agency that is reaching out to serve previously underserved populations, expanding services, and seeking new constituents by establishing itself as part of community development activities and as the city's prime ecology organization. The master plan demonstrated the new vision.[8] The training program gave greater status to employees. The tree planting and community forestry activities put the Department of Recreation and Parks in the neighborhoods, and Kids Grow and other education programs touched on the interests of all citizens.

4

The *social system* element sets demand on the *critical resources* and shapes how that demand is served and how resources are allocated. For planning and action research development, the URI team gave the most attention to four social institutions—*justice, education, leisure* and *governance*. *Justice* is about rules of behavior and standards of conduct that are shared by all citizens. The team placed emphasis on equitable access to the opportunities of civic life—regular garbage pickup, management of vermin habitat, open spaces for play, and leisure activities by all neighborhood residents. The conversion of vacant lots filled with garbage and rats was a step in that direction. Interns were determined and insistent until the city trucks and backhoes arrived and removed the trash. We have discussed *education* in terms of training and the efforts by Kids Grow and Outward Bound. The idea was that learning to grow a garden and keeping it up would reduce drug sales in the neighborhood and give the children and youth a stake and pride in their accomplishments and their neighborhood.

Baltimore's diversity of populations—which includes wealth, education, ethnic, religious, age, and other differences—means that diversity of *leisure* opportunities will need to connect interest in open space for all social class, ethnic, age, and other divisions. Seniors and youths might garden, wealthier and younger persons might bike on the new greenway, children might fish in the stream and use their skateboards on the greenway, seniors might go birdwatching and walk along the greenway. In order for the Gwynns Falls Greenway to attract visitors and serve as rehabilitation stimulus for the adjacent communities, it needed the involvement of police to provide a sense of security. It needed public works to repair the storm sewers that pumped raw sewage into the stream during heavy rainstorms. It needed professional advice from the city arborist and maintenance people in the Department of Recreation and Parks and the full participation of their Carey Murray Outdoor Education Center, which was at a midpoint of a completed greenway and uses the greenway as part of the outdoor teaching laboratory.

Two of the HEM social cycles—*institutional* and *environmental*—were most critical in accomplishing the research and management goals. Many of the public agencies involved in neighborhood development were at different stages of their life cycles. The Parks and People Foundation had been established more recently than the Department of Recreation and Parks. The Parks Department for the county was much younger than its city counterpart. Consequently, the internal norms, information structure, budgets, and mythologies, as well as adaptability and resiliency, varied considerably among these three partners. The environmental cycles were the cycles of flooding in the watersheds and factors that increase or decrease water velocities. The seasonal cycles of schools, sports, vacations, court sessions, marriages, and so forth affect demand on biophysical resources, volunteer labor availability, and social control efforts. For example, interns were driven by an academic cycle, but neighborhoods had more direct cycles of long and short term. The planting of certain tree species and shrubs did not match academic cycles. In the fourth year of the project URI requested and received salary funds to have some interns remain with the neighborhoods and therefore be on government and neighborhood cycles rather than academic cycles. This change helped the continuity and completion rate of projects initiated "off" season.

Age and *gender* were emphasized aspects of identity. Young people and seniors are two demographic groups with significant nonwork time and, therefore, made up most likely candidates for working on neighborhood greening projects. A high proportion of families in low-income neighborhoods tend to have women as the prime breadwinner. It seems likely (and has been borne out by experience) that there are many strong women in this cohort, most of whom want the neighborhood environment where their kids play to be improving rather than declining. In these social networks, *informal norms* are the effective controls since they are shared standards among locals rather than appeal to *formal norms* set and managed by outsiders such as laws, regulations, courts, and lawyers.

Differentials in *status* (prestige/respect) and *knowledge* of the street or the internal functioning of community ecology can be as useful as moving through the educational system as a way out of difficult situations. *Territory* becomes a critical means for sustaining neighborhood culture and internal stability in response to outsiders. The URI was to serve as the research and development arm of the Department of Recreation and Parks, producing knowledge about innovations and research findings on possible solutions to problems. In 1997, the National Science Foundation funded the Baltimore Long Term Ecological Research Study, which provided both basic knowledge and a systematic means for monitoring and evaluation measures.

Local gangs, government agencies, and development organizations have control over neighborhoods ("turf"), which they defend against other similar organizations. This mapping of social territories is critical for introducing new ideas such as community forestry.

The flows of *individuals, information,* and *capital* are familiar ones around the world. There is a rising tide of young single immigrants moving into the central, older cities and the newer postindustrial cities, a rising tide of middle-aged parents and dependent children moving to the suburbs. There is the flow of weapons, ideas, diseases, movies, terrorism, antiterrorism, drugs, and forms of information encapsulated in the fairly open exchange borders of the modern urban world. The domination of capital is evident in its flows and results, such as the financial crisis in Southeast Asia of the late 1990s; the rise of demand and price of oil in the twenty-first century as India, China, the United States, and Japan compete for limited supplies; and the global financial crisis and recession of 2008. These are large-scale trends that affect the context for the likely success or failure of revitalization efforts. Further, the demographic shifts in and out for city residents have real and direct consequences, as do the daily, weekly, and annual flow patterns of commuters, commodity suppliers, tourists, and sports fans. Flows of individuals, information, and capital from the state and federal governments become critical in the alteration of neighborhood development plans and support for programs, such as Revitalizing Baltimore, that supply the funds for urban community forestry activities and provide construction and planning funds for greenways such as Gwynns Falls.

5

The HEM helped the URI team to identify specific interventions, anticipate their interconnections, and estimate their likely contribution to URI goals. It served to organize and coordinate the efforts of many persons and organizations including state and federal agencies, local NGOs, neighborhood groups, interns, and professional and technical workers from the Department of Recreation and Parks. The URI had five broad goals: (1) to help restructure the Department of Recreation and Parks as a proactive expanded ecosystem services organization; (2) to use greening strategies as a means of revitalizing neighborhoods; (3) to reduce nonpoint source pollution of urban watersheds and the Chesapeake Bay; (4) to train multidisciplinary natural resource professionals to work in community development using an ecosystem approach; and (5) to gain experience with strategies and tactics that encourage and include ordinary citizens in becoming stewards of their community environments.

Work began with considerable cooperation from the Mayor's Office and Dr. Jones's replacement, Marlyn J. Perritt. However, as time went on, new challenges and constituencies increased, costs increased, and the tax base decreased. Under these conditions sacrifices were required, and one of the most easily sacrificed was the Department of Recreation and Parks, which was transferred to the Department of Public Works, whose director was eager to assert control over the lands and capital resources. However, in the course of time a new mayor, Stephanie Rawlings-Blake, was elected, and she restored much of the Department of Recreation and Parks to its independent status, but with greatly reduced personnel and fiscal resources and a new director whose emphasis was on efficiency rather than expanded services and organization. However, a new unit of community outreach was established, and this was influenced by the earlier community-based efforts.

Therefore, the actual hoped-for restructuring seemed more reflected in the Parks and People Foundation, which, under the leadership of Jacqueline M. Carrera, evinced dramatic organizational and programmatic change from where the URI began in 1989. The foundation assumed many of the functions once carried out by the Department of Recreation and Parks or that were planned to become functions of the "new" department—community forestry, including the Gwynns Falls Greenway promotion and connection to neighborhoods. A similar service was assigned to the Jones Falls watershed, whose trail was opened in 2012. URI interns and research and development, Kids Grow, reforestation and street tree establishment all continued to be active. Revitalizing Baltimore experienced some reduced funding under the Bush administration but still was central in coordinating work in the five-county area and the central city. The three watershed associations had been established and were building volunteer support. In September 2010, five watershed associations—Jones Falls, Herring Run, Gwynns Falls, Baltimore Harbor, and Baltimore Harbor Waterkeeper—legally merged into Blue Water Baltimore for "restoring, protecting, always." So the hoped for transformation that was originally sought did not come to pass, but many of the desired outcomes that were to flow from the restructuring are taking place but in different venues.

Many of the vacant lots converted to gardens and other green spaces have declined or returned to weeds. The original persons in charge became burned out, retired, or left. The larger hopes for greening and poverty reduction and improved educational results have not been fully realized. However, the foundation still provides mini-grants to neighborhood groups, and in some neighborhoods such as Pig Town and Sandtown-Winchester, where professionals and public money persist, there is continuing hope and clear restoration and revitalization as evidenced by the new street trees and ever-rising property values.

The important point is that many good results happened for the city and its people. These reflect the hopes and accomplishments of URI Baltimore and its partners. An annual report of Revitalizing Baltimore reveals that over its first 10 years:

> [It] has focused its efforts along stream valleys, parks and neighborhoods with significant tree deficits helping increase the tree canopy in 45 neighborhoods by planting over 7,000 street trees and 15,000 riparian trees and woody plants in over 650 projects involving more than 3,000 volunteer annually, and providing stewardship education to over 12,000 students and 900 adults. RB's twenty partnering organizations include the Maryland Department of Natural Resources, U.S. Forest Service, Baltimore City, Baltimore County, several nonprofit organizations, four watershed associations, businesses, and academic institutions. RB actively reaches out to culturally diverse communities to help residents plant trees along streets and streams, transform vacant lots to community green space, restore parks and schoolyards, and support youth education and adult training to foster stewardship of natural resources.[9]

The 2012 annual report of Blue Water Baltimore indicates a continuing set of accomplishments, having "planted 1,995 trees on 24.8 acres of land at 39 different planting events; Waterkeeper sampled 30 sites biweekly for bacteria, nutrients, and other pollutants; removed 61,460 pounds of trash at 78 cleanup projects; advocated to help pass 3 major environmental bills; installed 270 rain barrels and disconnected 44 additional down-spouts diverting over 330,000 gallons of storm water, painted 289 storm drains; installed 18 rain gardens that capture a total of 1.75 million gallons of storm water; our 4,355 volunteers served 10,616 hours."[10]

The Parks and People Foundation's 2012 Annual Report provided green grants to 450 community-based groups in the development and implementation of 771 beautification and greening projects throughout Baltimore City; helped transform 92 vacant lots into community garden spaces, neighborhood parks, and open green spaces; planted over 2,500 trees; created 91,896 square feet of green space; and recruited 359 volunteers who logged over 5,000 hours of work. They established a green career program to train youth for future jobs. They sustained the internship programs and community green development projects of URI. The Parks and People Foundation continued and expanded the Kids Grow and other educational programs where teachers are trained and kids get hands on education in the watersheds.

The anticipated agent of organizational change was not the Department of Recreation and Parks. However, the desired outputs and service seems more than substantial. The empirical reality of accomplishment is more important than keeping to the plan. Ideas were exchanged in a ferment of shared interests, and good people picked up on some of these ideas and made them their own and expanded the learning curve for all. The leadership over the years by Dr. Morgan Grove, an urban research forester with the U.S. Forest Service, has been critical, and the warmth and generosity of many collaborators have given the participants hope when despair might have been the reality. Although the ideas came from many sources, the HEM was essential to the beginning of the effort and continues as a catalyst of intellectual legitimacy for the idea that one can apply concepts and methods of ecology to understand and solve urban problems.

6

A consistent finding of many urban studies by historians, sociologists, engineers, anthropologists, political scientists, artists, and others is that the city is not understood as an island apart from the whole; rather, it must be seen as a part of a larger temporal and spatial context within which it resides. We must also recognize the necessary importance of the plumbing of the place—the anatomy of providing potable water, sewers, garbage production and removal, behavior of watersheds, storm run-off, markets, and life and death rates of different population segments.

The work in Baltimore demonstrates that an ecosystem approach (HEM) has considerable utility for: identifying basic elements and their influence on the structure and function of an urban system; anticipating likely connections in the elements of the system; providing the means to organize data collected and to capture the lessons learned as the consequences of various policy or management actions; designing more effective and efficient performance-based public services; and improving our ability to more effectively monitor and evaluate various ecosystem interventions and actions. Overall, the approach permits a more efficient, effective, and equitable allocation of scarce resources for urban challenges by considering how policy interventions intended for one issue cascade in a systematic way through other issues, agencies, and subsystems. The next chapter will continue to focus on the Baltimore ecosystem and examines research as a complement and guide for the policy, planning, and management of human ecosystems.

T·E·N

Toward a More Perfect Civic Order
Lessons Learned from Research

1

The previous chapter reported the results from applying the HEM to the URI in Baltimore. However, it is not just a story about Baltimore. Rather, it is a working example of the general awakening of conservation organizations throughout the world. This awakening affects ecosystems from wildlands to urban park lands. In this transition, organizations are learning that if they hope to sustain biophysical systems, they need to pay more attention to how their actions affect the health and interests of local human communities.

This transition is still a work in progress as new professional educational curricula are put in place, as changes in the recruitment of professionals are made, and as changes in the cultures of the organizations are established. The Baltimore case highlights how the lessons learned from using the HEM can be part of the transition. Resource professionals accustomed to working in the lightly populated corners of the world can gain necessary lessons emerging from urban places on what works and what does not as they try to integrate human realities into their policies, planning, and management practices. Urban ecosystems represent more concentrated places where, most often, neighborhoods resemble villages jammed together spatially; they are culturally separate yet share universal properties of behavior found in all small-scale communities.

The benefit in urban ecosystems is that these understandings of the properties of behavior can be based on data that have broader and longer time frames than are available for most rural villages.

Our understanding of human ecosystems comes from some mixture of personal experience and research. Personal experience gives a sense of the importance of small details, while research adds observations that are systematic, measurable, and reproducible. If we want to know about survival threats for elephants or wolves or turkeys, we need to study their social behavior. If we want to stop poaching of elephant ivory we need to study characteristics of trade networks, the rate that ivory can be poached, the livelihoods of local people, and the motives of the ultimate consumers. This critical knowledge is similar to the need for understanding present and future trends of supply and demand of other natural resources, such as those provided by wildland or urban ecosystems.

The HEM provides one means for linking the biophysical and sociocultural elements in conservation decisions. Our notion of a human ecosystem is one of unified attention to co-variation between key variables. Urban ecosystems compress and expand these patterns of understanding. This is particularly so when we consider the green spaces managed by public and private conservation organizations. In Baltimore, the major units of analysis were the four watersheds—Jones Falls, Gwynns Falls, Herring Run, and the Inner Harbor. The subunits were the self-defined neighborhood communities. Those working in non-urban areas can use these lessons for creating similar kinds of measures and understanding.

We will consider specific HEM lessons learned from a wide range of research on critical issues facing organizations like the Baltimore Department of Recreation and Parks and its partner, the Parks and People Foundation. Both organizations exist within conditions where funding and personnel have for many years been restricted. Like most park and conservation departments around the world, their limited fiscal situation has been made worse by the cumulative postponement of many necessary activities and the lack of maintenance, both of which expand the problem. So the issues are many, and the fiscal support is limited. Further, unlike most federal and state conservation agencies, city park agencies seldom have their own research divisions. Consequently, they have difficulty in anticipating trends in the characteristics of likely future visitor populations, future public service needs, and opportunities for meeting those needs. Further, they do not have a means for systematically monitoring and evaluating past, present, and future programs. Under these conditions the transition of conservation organizations to better serving communities is more a boom-or-bust operation than a more sustained effort.

In Baltimore, the URI and its partners attempted to fill the research void and reduce some of the boom-or-bust dimensions. The work was not traditional science in the sense of chemistry or physics or publishing in academic science journals but rather a research and development effort to solve practical needs challenging local conservation efforts. The more abstract scientific research of the Long Term Ecological Research (LTER) in 1998 could build on a platform of utility from this early work. The early work gave answers to immediate problems rather than the advance of a specific scientific discipline.

We will examine a varied set of research studies that were undertaken to illustrate how lessons were learned. These examples offer lessons as to how a range of research types, methods, and approaches can help conservation organizations to include human behavior as a natural part of the ecosystems being managed. Our first examples come from work done by graduate student interns. Their research was one part of their intern service and a further extension of their training. The second set of examples reports on applied work done in cooperation with the Parks and People Foundation and the Department of Recreation and Parks on creating an urban greenway system. The third example illustrates how the HEM served in the design of the research on visitor needs in national park units (the agency term for individual parks) that had findings that were very useful for managing city parks. The studies were done on visitor trends and needs comparing three nearby national park units, with core attention on the Fort McHenry National Monument and Historic Shrine within the city of Baltimore.

The fourth set of HEM-directed research cases are given the most elaboration in their findings because the studies and their data clearly suggest some universal empirical patterns found in nature-society interactions. Also, the data presented give an idea of the utility and value of such work. This section is based on a policy research report done as a class project on the application of the HEM. It was done by a group of three graduate students from different environmental disciplines in 1999. This student work was expanded to a time series that illustrated how different authors and research efforts can complement one another through use of the HEM; in this case, the student work connected to a community-based data set on indicators of vital signs of community social and economic health. We provide findings from these studies to illustrate the utility of the HEM in linking information from different research efforts to gain a common timeline on community responses to various internal and external events.

What follows is a short discussion of the National Science Foundation (NSF) and U.S. Forest Service (USFS)–funded Long Term Ecological Research (LTER) that has been going on in Baltimore since 1998. This research is in its third cycle of renewal and was the first such study to examine the urban ecosystem as a specific habitat. Though it is tied to community service, its primary goal is contributing to ecological science knowledge. The multidiscipline work is organized around the HEM.

2

We begin with some of the early efforts at building a cooperative research base that had interns joining with local communities, public natural resource professionals at the state and municipal levels, and the innovative responses of individual students. The first Baltimore URI report set the attitude and directions that would guide the future efforts of the URI program.[1] The participants in the intern program were persons who had mostly experienced cross-discipline research in theory and talk as a "good," even necessary, thing to do. Now they were in the field directly experiencing the difficulties of applied, participatory cross-discipline research. It is these learners who will be the teachers of the next set of conservationists. So the research is directly useful to the agency, but it is equally important in building more effective and equitable natural resource professional service.

The first URI report describes the methods and guidelines for research on feral animals in large urban parks. In this case Gwynns Falls/ Leakin Park was also the venue for research on the management of invasive vines. This research involved students from a youth training effort who worked on collecting the baseline data and doing experiments with vegetation management controls. Other studies were on the utility of GIS for park management—remember, this was 1990, and the technology was just emerging, and the interns were the innovative carriers of its benefits. The studies on community forestry techniques were developing and testing formal and informal means for giving solid experience in street tree management with a training manual—"Community Forestry Stewardship Handbook"—and field training of neighborhood people along the watershed gradient.[2]

None of this research work was easy, as there were no traditional guidelines on how to combine local knowledge and experience with the multiple disciplines necessary to understand the system. As the project leader's report notes: "One of the URI's main objectives is to research the practical aspects of sustainable resource management. This goal will remain unattainable if the natural and social sciences do not seek ways of incorporating humans as ecological factors in hypotheses and paradigms. The roots of the human/nature split are deep. Healing will require interdisciplinary education programs, creative scientific modelling and further research of urban ecosystems."[3]

Most of the student intern research focused on relations between HEM biophysical variables, such as plants and animals, and educational and leisure institutions and the influence of demographic and environmental cycles. The work from these studies and those that followed gained from the HEM in two ways. Its conceptual categories compel the researcher to consider which connections between people and nature are critical to understanding their given research problem. Further, it is a "market basket" for storing present data sets to accumulate toward general trends and probable connections between elements in the system. That is, enough information can be sorted to give an idea of what probably works and what probably does not under certain conditions. The HEM does not give absolute predictions about relations between system elements but accumulates and stores findings so that the researcher or manager gains better probabilities of likely outcomes. The HEM does help in avoiding mistakes made in the past and suggests better options for the future with a better chance of making more effective and sustainable decisions. The combination of student and citizen science efforts gives quality efforts at reasonable costs and expanded skills for being better stewards of human ecosystems.

3

The second research example highlights the versatility and utility of the HEM for cooperative research that serves many different communities and visitors.[4] This research was directly related to a specific planning and management issue—the establishment and design of a multiple-use greenway for the Gwynns Falls watershed. This research had more time and resources to begin and close its work than did the student intern efforts. The work also provided guidelines for similar greenways on the Jones Falls and Herring Run watersheds. This study described, mapped, and interpreted the structure and process of the social ecology of Gwynns Falls Park. Systematic data came from a variety of sources: systematic observation, census tract data, historic sources such as the Olmsted brothers' plan (1904), and informal interviews with local citizens. This study mixed maps, text, and statistical tables to define the nature of the adjacent neighborhoods and the likely demands, constraints, and opportunities that the greenway and its neighborhoods would need to consider. The study specifically identified points of threat to the security of visitors and nearby residences from regular points of trash dumping and attendant rat colonies, drug dealing and drug use, and prostitution enclaves.

In 1999, a much more detailed set of maps and data was produced for the Gwynns Falls Watershed Association by the Revitalizing Baltimore Project, which not only served the planners and managers but also served to empower local people in the effort to restore the park and expand its services to local communities and the city at large.[5] Neighborhoods used the information to create access trails to the greenway and to develop means for protecting recreation sites from illegal activities and security threats. The LTER project made the Gwynns Falls watershed the base for its scientific research efforts.

4

The third example is a study on perceptions and use of the Fort McHenry National Monument and Historic Shrine in Baltimore. This study was done in 1991 with a modest grant from the National Park Service and compared a sample of Baltimore metro region visitors to this park with visitors to three other nearby NPS units—Assateague, Shenandoah, and Gettysburg. The authors hoped that the findings of the research would aid the park planners and managers in serving local and national visitors and in sustaining the biological and cultural resources of the park.

One of the research goals was to help Baltimore parks managers understand how their work in individual parks was connected to larger ecosystems and how their efforts could link with the region's many open space opportunities. In a sense, Fort McHenry is an anchor to the Gwynns Falls Park and the Gwynns Falls Greenway and an important tourism gateway for the city. The fort was part of the harbor development strategy. The park's location within a mostly white, working-class residential community was a unique opportunity for both the park managers and the local residents to understand each other's perceptions on the role of parks in the human ecosystem. Often these perceptions were very different.

Little comparative research had been conducted on national park units in the United States that exist in urban ecosystems. Fort McHenry National Monument and Historic Shrine, the base park for this project, was established in 1925. It is situated at the tip of a peninsula between the northwest and middle branches of the Patapsco River. It is within easy reach for people in the local neighborhoods and is a component of Baltimore's Harbor renewal plan. In 1990 there were 561,000 visitors to the park unit, and they collectively spent 842,000 visitor hours at the park site. Nearly a majority of the visitors (48%) estimated that they had spent $25 or more associated with the visit.[6]

The visitor population can arrive by a wide range of transportation modes—walking, bicycling, private automobile, municipal bus, taxi, and water taxi. So our onsite interviews had to adapt to these realities under the guidance of park employees. Further, there was a larger regional population that may not have visited the Fort McHenry site; for understanding these visitors the research team worked with a professional opinion research organization that had long-term experience doing market research in the region. They conducted a telephone survey of Baltimore metro residents using a standard form that was developed by the research team with the advice of the research organization professionals. Another visitor population group was schoolchildren, who came in school buses with their teachers. The teachers were interviewed about their experiences.

The Baltimore Department of Recreation and Parks Strategic Action Plan incorporated some of the findings from the study. There were eight reports on the field interviews published by the URI in 1991 and 1992; in 1995 the report on schoolchildren visits was published, and three reports on the telephone survey data came out. All of these studies were guided by the human ecosystem approach. The park unit may have specific legal boundaries, but the research examined it not as an isolated unit but rather as an integral part of a larger biophysical and sociocultural system. In this case it was part of the greater Patapsco River and Baltimore Harbor ecosystems.

We highlight some of the findings as they reflect the HEM application. Note that several biophysical and cultural resources were a major part of the analysis. Air, water, and land were critical determinants as to why the fort was located where it is. For visitors it provides some of the most outstanding views in the city with the river, the boats, the harbor, and its urban background. As Francis Scott Key could confirm, it is a most elegant place from which to develop a national anthem. Cultural resources such as myth and art enhance these biophysical elements, while social institutions such as family and friends, leisure and education are reinforced by social norms.

The research documented that visitors were seldom a single person who came alone and most frequently came as members of families and/ or friends (65%). Learning from the park's human and natural history is cycled through these social networks in very different ways than a single individual alone who does not have the immediate reinforcement and retention found in a group. This is important as the management of behavior is mostly done within these social circles. Further, the teachable network is dependent on these reciprocal relationships. This may seem obvious, but for most park managers the visitors come as individuals and information programs are developed to serve the individual rather than the actual user group that is there. In a sense the park is a means rather than an end point. It serves as a place and a rationale for people to be together with those they most care about. This is information that can and should enable parks programs to serve those who come to their programs.

The HEM helped to bridge information levels between different components of the ecosystem. The management agency realized that by just looking inward to the park property it was missing many of the forces determining demand and onsite behavior. The agency gained a means for understanding how families and other significant elements of the ecosystem interact with one another and how the watershed boundaries and the neighborhoods that are strung along them are part of the influence on recreation behavior—positive or negative. With this structure, an agency has a means for developing management strategies and tactics that better fit the empirical reality of parklands. For example, encouraging neighborhood people to work together to build trail access to their part of the Gwynns Falls watershed not only builds a trail for access but also builds a sense of community stewardship of the natural system and instills the norms of appreciative park behavior.

5

In 1999, there were graduate student class projects that required three student teams to write on particular HEM applications for open space issues in Baltimore. We will use one particular student study to stand for most of the past, present, and future student efforts. The student research team developed analyses of community trends by examining seven neighborhoods that represented a wide range of variation, from those with a low median income to those with higher incomes as well as a wide spectrum of racial composition. However, our analysis here will consider only their work on two communities—Homeland and Madison-Eastend, which represent extremes in terms of income and social status. We will then use material by a community participation group, the Baltimore Neighborhood Indicators Alliance (BNIA), which developed sets of quantitative indicators on these two neighborhoods as well as all other neighborhoods and the entire city. These indicators are called Vital Signs and are reported annually on the BNIA website by community, topic, and indicator. Past reports are archived. We will look at the same two neighborhoods and the entire city with these measures of our neighborhood's health and strength over time. We are most interested in demonstrating how the tactics of using available data regrouped by the HEM and enhanced by qualitative data sets can provide a useful understanding of complex human ecosystems.

The student researchers had two short, intensive field opportunities, with the bulk of their data and observations coming from published or internet available resources. The BNIA used similar methods for a "community driven process," focus groups, key informant interviews, and other techniques to "hold one another accountable for our efforts to rejuvenate the city."[7] The work is unique as it works at the neighborhood level. They developed forty "vital signs" that were grouped into eight topic areas: housing and community development, children and family health, safety and well-being, workforce and economic development, sanitation, urban environment and transit, education and youth, and neighborhood action and sense of community. Their data sources cover a wide array of information, from census data to reports by private organizations.

The BNIA quantitative data were combined with the data and observations of the student team to identify trends in key indicators of the well-being of the human ecosystem. This combination of different data sources illustrates the utility of the HEM as a link between the qualitative work of the students and the quantitative work of the BNIA. This combination of data extends the time trend for understanding the structure and function of "social patches" within the larger urban ecosystem, gives a human feel for the variation between neighborhoods, and provides a richer reality when we monitor and evaluate interventions that seek to develop strong neighborhoods, good quality of life, and a thriving city over time.

Our example reports on two of the graduate student neighborhood profiles, Madison-Eastend and Homeland. As noted above, we extended their observations with a re-construction of the BNIA vital signs data into HEM categories for these two neighborhoods and the entire Baltimore city data. The reader can browse the attached tables to learn some of the available data and how the HEM can group these data for measuring the impact of natural resource decisions.[8]

Table 10.1 provides information on how the research team developed some of its indicator measures. Most of the census-based information is clear. However, what the team did was use census tract units and relate them to specific neighborhoods. The geographic units are the reference point. If the table identifies Madison-Eastend, it is that specific population within that geographic boundary. Or, when the table identifies a percentage of persons are black or elderly, it is bound to that statistical unit's total population.

We provide a shortened version of the BNIA reports to illustrate the possibilities in using available resources for applying the HEM. The lessons to be learned are not in the specifics of these data comparisons but in demonstrating that there is a good deal of overlooked data in the sources and that there is much value in re-analyzing available public data from census studies, market research studies, national park studies, and local research resources like the BNIA. There will never be sufficient funds for the social side of human ecosystem studies, so the researcher needs to be creative in uncovering these potential research resources. The 1999 student profile study demonstrates how a group of creative scholars with limited time can provide a useful human ecosystem study. The community-based BNIA data sets help to extend that earlier analysis, and these two data sets have inspired the later work of the NSF Baltimore Ecosystem Study.

The strength of this effort is the model that provides abstract concepts for organizing a functional set of observations that are repeated in routine fashion for humans wherever they may live. These conceptual insights are tactically tested in the ability to compare case events within and between similar situations to see which variables are most crucial for driving other variables in the biosocial system. The reader should be aware that this information is reporting on events at that time. A visitor today would find many persons less hopeful and many associations no longer active. Further, there are many new projects and activities that were not in play in 1999 but are now emerging as useful possibilities.

Table 10.1. Components, variables, measures, and sources of data developed for the Baltimore community research

Component	Variable	Measure	Source	Comments
Biophysical Resources	Land	The number of residential properties located within an area identified by Maryland Property View.		It is important to note that this indicator is a count of properties (single family homes, condominiums, and duplexes) and that a property can be composed of multiple units. There are other land use measures that we could have used, such as amount of commercial properties and so on. Residential properties' proportion of land use seemed a good clue as to the predominant behavioral patterns that would likely occur and differentially affect natural systems such as stream run-off with more permeable soils or directed flows of rain water from roofs and parking areas.

(*continued*)

Table 10.1. (*Continued*)

Component	Variable	Measure	Source	Comments
	Energy	Percentage of commuters aged 16 and above that walks to work.	American Community Survey, 2008–12	There were several other indicators on commuting, use of water, and so on. But this indicator seemed most salient for comparing trends and between different neighborhoods.
	Flora	Percentage of total land area comprised of tree canopy in an area (within the designated community borders)	The primary sources for this land cover layer were 2004 pan-sharpened 1m Ikonos satellite imagery, a normalized Digital Surface Model (nDSM) derived from 2006 LiDAR; data and LIDAR intensity data resulting from the 2006 acquisition.	Other sources of data include the City's planimetric GIS database (building footprints and road casing polygons). The land cover classification was performed using automated object-based image analysis (OBIA) techniques in Definiens Developer/eCognition Server. No accuracy assessment was conducted, but the data set was thoroughly reviewed at a scale of 1:2000. Over 370 corrections were made to the classification. Analysis by: University of Vermont Spatial Analysis Lab. Data current as of 2007."

Socioeconomic Resources	Information	The percentage of persons aged 16–19 who are in school and/or are employed out of all persons in their age cohort.	American Community Survey, 2008–12	Measures youth participation in productive activities.
	Capital	The percentage of households that pay more than 30% of their total household income on mortgage and other housing-related expenses.	American Community Survey, 2008–12	Measures housing burden.
		The percentage of households that pay more than 30% of their household income on rent and related expenses out of all the households in an area.	American Community Survey, 2008–12	Measures housing burden.

(*continued*)

Table 10.1. (*Continued*)

Component	Variable	Measure	Source	Comments
	Labor	The number of persons between the ages of 16 and 64 not working out of all persons, not just those in the labor force (persons who may be looking for work).	American Community Survey, 2008–12	These persons are seeking work that pays a formal income.
Social Institutions	Justice	The rate of service requests for dirty streets and alleys through Baltimore's 311 system per 1,000 residents.	Baltimore City CitiStat. 2010, 2011, 2012; U.S. Census, 2010	More than one service request may be made for the same issue but is logged as a unique request.
	Justice	The rate of service requests for addressing clogged storm drains made through Baltimore's 311 system per 1,000 residents.	Baltimore City CitiStat. 2010, 2011, 2012; U.S. Census, 2010	More than one service request may be made for the same issue but is logged as a unique request.

Justice	The number of persons aged 10 to 17 arrested for violent offenses per 1,000 juveniles that live in an area.	Baltimore City Police Department, 2011; U.S. Census, 2010	Violent offenses may include homicide, rape, assault (with or without a weapon), and robbery. This indicator is calculated by where the arrested juvenile was arrested and not by where the crime was committed. Arrests are used instead of crimes committed, since not all juveniles that are arrested are charged with committing a crime. This indicator also excludes offenders who are later charged as adults for their crime(s).
Family	The rate of calls to emergency 911 for domestic violence per 1,000 residents in an area.	Baltimore City Police Department, 2010, 2011; U.S. Census, 2010.	Calls for service are used rather than actual crime incidence since domestic violence can be classified as one of several types of criminal offences. It is important to also note that not every case of domestic violence is reported and some claims of abuse may be unfounded.

(*continued*)

Table 10.1. (*Continued*)

Component	Variable	Measure	Source	Comments
	Health	The percentage of births where the mother received prenatal care during the first trimester of the pregnancy in a calendar year out of all births within an area.	This information is calculated by the Vital Statistics registration information collected from each live birth, 2011, 2012	
	Governance	The percentage of persons who voted in the last general election out of all registered voters.	Baltimore City Board of Elections, 2010, 2012	A proxy measure designed to reflect neighborhood action and participation in community life.
	Knowledge	The number of children who have registered for and attend (1–5; 6–8; 9–12 grades) as of September 30.	Baltimore City Public Schools, 2009–10, 2010–11, 2011–12	This count only includes students enrolled in public schools.

	The percentages of students passing M.S.A. exams in reading and mathematics (assorted grades).		Maryland School Assessment (MSA) scores measure the number of students scoring in one of three classifications out of all students enrolled in that grade. Students can either be rated as advanced, proficient, or having basic knowledge of a subject. This indicator includes only those students who have tested as advanced or proficient.
Social Cycles	The median number of days that homes listed for sale sits on the public market in a given area.	RBIntel. 2010, 2011, 2012	This time period is from the date it is listed for sale to the day the contract is signed. Private (non-listed) home sale transactions are not included in this indicator. The median days on marked is used as opposed to the average so that both extremely high and extremely low days on the market do not distort the length of time for which homes are listed on the market.

The dynamics of urban ecosystems do not wait for lessons from policy experiments to emerge but respond to the next new challenge.

To the social ecologist, these observed land use clusters are similar to what the biological ecologists call the patch system, where expected groupings of plants and animals occur and a certain level of perturbation serves to sustain the biodiversity of the larger system. These patches in human- and nonhuman-dominated systems provide an economy of use and expectation. The nonhuman ecologist can explain the "causes" of these observed regularities as due to soils or climate or recent perturbations, whereas the human ecologist can note the regularities but is less able to derive causal elements. So for the human ecologist patch analysis provides a useful means of grouping observations but cannot assign causal connections in the comfortable and confident way of the biological ecologist.

Treating these neighborhoods as specific patches integrated with the larger urban ecosystem serves as the context for interpreting findings collected by the students and BNIA. We begin with a paraphrase of the student's description of the Madison-Eastend neighborhood in eastern Baltimore at the time of the project. It was bounded by Milton Avenue, Eager Street, Edison Highway, and Monument Street. The north side of the neighborhood was bordered by a railroad with a park on the east side. Johns Hopkins University was a few blocks to the west of the neighborhood. It is a flat part of the city, with the streets laid out in a grid system. In 1999 most of the homes in the area were two-story row houses mostly built during the 1930s. Few to no houses had front yards, back yards were often concrete, and the blocks were visually monotonous. At the time of the project, the average value of these houses was among the lowest in Baltimore, and only 41 percent of the houses were owner-occupied. However, the median rent value at the time, $475, was slightly higher than the city average, and 34 percent of households were receiving some type of public assistance. The population density here was far higher than any other location studied in the project, with 43,721 people per square mile. Madison-Eastend had a young population, with 38 percent of people under 19 years old. The birth rate to teenage mothers was far higher than other neighborhoods in the project. Unmarried females headed over half of the families.

Overall education rates were low. Table 10.2 presents selected data on the Madison-Eastend community.

The researchers used only the most necessary parts of the HEM to guide their selection and use of available information. There was a mixture of quantitative and qualitative data sets so that the reader of their report would gain an understanding of the challenges to the neighborhood of a large dependent population and the characteristics of the grid\housing system that create a certain aesthetic environment. The researchers' tactic was to use "hard" numbers in census and historic public documents to gain an understanding of the constraints and opportunities placed on the local people. They then fleshed this description out with key informant interviews, systematic observation, and some in-depth formal interviews with youths and community leaders. These data sets permitted them to make intra-group comparisons as to changes that sustain or alter the biosocial factors of the neighborhood.

Homeland was the second community the project focused on. It is in the northern part of the city and shared the upper-class characteristics of adjacent neighborhoods like Guilford and Roland Park. Homeland lacked the gridded nature of the southern Baltimore neighborhoods, with winding, gently sloping streets and cul-de-sacs. The neighborhood was built around a central island of public open space, with two small ponds with fountains.

The neighborhood was formed from a large agricultural estate in the middle of the nineteenth century. At the time of the project, Homeland was a wealthy residential neighborhood characterized by relatively large houses, low housing density, and areas of open green space. It was reminiscent of a suburban county community with its natural, flowing topography, tree-lined streets, and lack of city business and street life. Like other wealthy areas of the city at the time, there was a relatively high elderly population, which was balanced by an equal representation of school-age children. Almost 90 percent of the families were married couples, although there were a few single mothers. The area featured low rates of births to mothers under 17 years old. Table 10.3 presents selected data on the Homeland community.

Although Homeland is only a fifteen-minute drive from Madison-Eastend, its social profile suggested a very different world

Table 10.2. Madison-Eastend and the Human Ecosystem Model

Critical Resources		Human Social Systems	
Natural		Social Institutions	
Land	• High housing density • Few street trees, yards	Commerce	• Close to downtown • Few local businesses • Drug trade
		Education	• Rose Street classes
		Social	• BES, Johns Hopkins HEBAC involvement
Socioeconomic		Social Order	
Population	• High density	Age	• 38% under 19 years old
Labor	• High unemployment	Class	• Low income
Capital	• Low income per capita • Low housing value • School scores low • Few with professional degrees	Ethnicity	• 77.6% black
		Clan	• 35% married families, • 55% female households
		Formal	• High crime rates
		Informal	• High teen births
		Territory	• Lower owner occupancy
		Knowledge	• Few with professional degrees

Cultural

Organizations
- Rose Street community center

Beliefs
- Unification against violence and trash
- Black pride increasing
- Local Sunday sermons

Social Cycles

Patterns
- Weekly sermons and trash pick-up
- Seasonal garden plots
- School cycles

Hypothetical Links

1) Poverty and unemployment may make this community susceptible to drug trade violence, and fear of violence may isolate neighbors.

2) Currently developing cultural resources at the Rose Street Center may show potential for increased socioeconomic and natural resources over time.

3) Low income levels, high teen birth rates, a young population, low homeownership, and low investment in the neighborhoods (yards, housing values) suggest high levels of motivation for change.

Table 10.3. Homeland and the Human Ecosystem Framework

Critical Resources		Human Social Systems	
Natural		Social Institutions	
Land	• Low population density • Predominantly residential • Original topographic features • Managed for aesthetics	Education	• High number of schools (particularly private)
Socioeconomic		Social Order	
Capital	• High income	Age	• Evenly distributed population
Information	• Easy access to information	Class	• High income, high degree attainment
Education	• High percentage of graduate/professional degree	Informal	• Traditional family structure • Cultural mores; low births below age 17, low # single mothers • High education standards
		Territory	• Large yards, access to open space

Cultural

Organizations
- Neighborhood group
- Umbrella community association
- High number of schools and churches

Beliefs
- Equate aesthetics with a perception of health
- Land equals money

Social Cycles

Individual
- Daily commuter flows

Hypothetical Links

1) Importance of privatization: Implications that private management of education and open space lead to better results. This cannot be considered without focusing on high-income/white-collar nature of the neighborhood.

2) Commuter community keeps enterprise and development out of their own community. This may be an issue of power related to education level, age structure, and income distribution. This is connected to the idea that money and power allows greater beauty. Compare Homeland's natural topographic features and highly managed green areas to lower-income/commercially developed areas.

3) Strong informal norms influence the formation of factors such as aesthetics and education level.

with different life chances. The demographics suggested a rapid turnover in the near future as the elderly white resident population would soon decline and the neighborhood would shift to a different ethnic and demographic mix.

Tables 10.4 and 10.5 permit the comparison of the indicator data restructured into HEM categories. Where it seemed useful, we do include some 2010–11 data. Still, our interest is in trends, not the absolutes of the moment. Also, the Homeland data are grouped with data from two other neighborhoods, so there is some deviation from the 1999 study. However, there are temporal trends that permit us to ask some better questions about what is going on. For example, the residential properties for the city of Baltimore and the Homeland community saw a slight increase, whereas there was a slight decline in Madison-Eastend. It is interesting that a higher percentage of people walk or bike to work in Homeland (a commuter neighborhood) than Madison or the City; further, tree canopy is higher in Homeland than in the City, and there is hardly any such thing in Madison. A careful reader might begin to wonder about the relationship between greater greening in a neighborhood and quality of life. Some of these associations are reported in research done by the LTER studies. These studies do ask about the "cause" or "consequence" of income differences and greening along with other measured differences. New questions and discoveries emerge from the HEM grouping of information for researchers.

If we look at capital creation, there were much higher proportions of household income paying off mortgages in Madison than in Homeland, and a higher proportion of Homeland households were in owner-occupied homes. The stress of trying to pay off one's loans was greater on the smaller number of families owning their homes in Madison than it was in Homeland. Mortgage policies press on this fragile level of middle-class hope and most likely will push many families over the brink. This was less likely to be the source of stress in Homeland. These comparisons help one to see how the resources, norms, and institutions interact in such a way to keep a place in despair. The critical points for interventions that must go on for many years to actually gain traction for development and relief for the family lives can be identified. We have added data tables from the student project with a fuller set of data to

Table 10.4. Critical resources*

	Madison-Eastend	Homeland	Baltimore City
Biophysical			
Land (residential properties)	2000 = 3608 2001 = 3601 2002 = 3586 2003 = 3582	2000 = 3623 2001 = 3740 2002 = 3744 2003 = 3732	199,013 (2000) 202,309 (2011)
Energy (% walking work)	13	15.7	6.7
Air (day above 90°)	NA	NA	2000 = 4 2001 = 10 2002 = 16 2003 = 2
Flora (% tree canopy)	1.4	49.5	27.4
Socioeconomic			
Information (% work/school)	76.6	100.0	87.4
Population (% black)	90.3	11.8	63.8
Capital (% *own* pay 30% month)	35.7	16.5	31.6
Labor (% pop. 16–64 unemployed)	11.7	3.7	7.9
Cultural			
Organization (# associations)	2003 = 16	2003 = 24	2003 = 650

*For some of the data where comparative base is not clear we have listed this information in Table 9.x at the end of these data tables.

Table 10.5. Social system

	Madison-Eastend	Homeland	Baltimore City
Institutions			
Family (domestic violence)	2000 = 109.6 2001 = 87.2 2002 = 76.0 2003 = 58.2	2000 = 10.8 2001 = 13.3 2002 = 12.3 2003 = 6.7	2000 = 57.1 2001 = 58.7 2002 = 54.3 2003 = 43.7

(*continued*)

Table 10.5. (*Continued*)

	Madison-Eastend	Homeland	Baltimore City
Health (maternal care)	2000 = 56.1 2001 = 52.0 2002 = 58.4 2003 = 65.9	2000 = 94.2 2001 = 90.2 2002 = 93.5 2003 = 96.2	2000 = 72.1 2001 = 72.1 2002 = 76.5 2003 = 74.9
Justice (pollution index)	2002 = 12.8 2003 = 20.2	2002 = 2.1 2003 = 9.6	2002 = 7.7 2003 = 14.9
Commerce (% owner occup.)	2000 = 43.2 2001 = 40.0 2002 = 39.5 2003 = 37.0	2000 = 85.9 2001 = 85.7 2002 = 85.9 2003 = 75.4	2000 = 65.8 2001 = 64.6 2002 = 64.1 2003 = 62.9
Governance (% voted)	2010 = 30.6	2010 = 59	2010 = 44.4
Education (dropout rate)	10.3	7.1	11.7
Social Cycles			
Individual (pop. cycles by age [%])	0–17 = 36.7 18–24 = 9.6 25–44 = 28.8 45–64 = 18.6 65+ = 6.3	0–17 = 13.5 18–24 = 27.4 25–44 = 22.8 45–64 = 21.8 65+ = 14.5	0–17 = 24.7 18–24 = 10.9 25–44 = 29.9 45–64 = 21.2 65+ = 13.2
Institutional (days prop. On market)	2000 = 77 2001 = 45 2002 = 35 2003 = 25	2000 = 13 2001 = 13 2002 = 13 2003 = 8	2000 = 52 2001 = 46 2002 = 36 2003 = 28
Social Norms			
Population	8,929	16,910	651,154
Gender (demog.)	Male = 4,117 Female = 4,812	Male = 7,532 Female = 9,378	Male = 303,687 Female = 347,467
Class (hsehld. cash)	$26,460	$58,995	$30,078
Status	56.0	87.5	64.4
Advanced programs			
Knowledge (enrollment)	1–5 = 827 6–8 = 552 9–12 = 551	1–5 = 141 6–8 = 98 9–12 = 116	1–5 = 34,613 6–8 = 22,007 9–12 = 25,576
Territory (illegal dumping)	2002 = 23.0 2003 = 13.4	2002 = 1.6 2003 = 0.8	2002 = 9.5 2003 = 5.4

encourage the reader to look more closely at the empirical possibilities, opportunities, and internal and external constraints on these communities. The value of the HEM is that it can help to tease out the behavioral regularities and to describe the characteristics of structural patterns and the variations in the dynamics of urban systems as we move across the social spectrum.

The lessons learned from these tables is that three graduate students on a term project and a group of local neighborhood persons can join forces at relatively low cost to produce critical information for monitoring and evaluating trends in specific human ecosystems. Most resource professionals want to know what works, what does not, and why. These are examples to guide that kind of work.

There are several comparisons worthy of consideration from these tables. The figure on tree canopy (table 10.4) has an astounding rate, as already noted. With 1.4 percent for Madison and 27.4 percent for the city at large and 49.5 percent for Homeland, the forest canopy is a major factor in establishing an image of life quality. These environmental factors take on greater depth when we consider the service requests related to illegal dumping (table 10.5) at 23.0 and 13.4 for Madison but only 1.6 and 0.8 for Homeland. In short, like most U.S. cities, Baltimore has low-income, struggling neighborhoods that not only suffer high unemployment and low educational attainment but also must put up with daily visual reminders of their diminished status. Additional data are presented for the two communities and the city of Baltimore in tables 10.6, 10.7, 10.8; they are presented here to provide suggestions and insight for others who wish to use the HEM to understand human ecosystems.

We think these examples give the reader a good introduction of what bright and determined students and a group of citizens can accomplish in seeking to understand and measure changes in neighborhood patches and the impact of rehabilitation policies on those patches. Though this information is from a city, the tactics and techniques can be replicated for non-urban venues. It also demonstrates the utility of the HEM in grouping and connecting different data sets. These very applied efforts demonstrate an HEM-based framework that the ecological sciences can build on in their ventures into the unfamiliar urban habitat.

Table 10.6. Madison-Eastend: social profile (data from sources other than census)

Family Cycles	2000	2001	2002	2003	Children and Family Health, Safety, and Well-Being	2000	2001	2002	2003
Domestic violence rate	109.6	87.2	76.0	58.2	Teen birth rate	158.8	130.3	106.6	128.0
Child abuse rate	21.3	17.1	NA	NA	Maternal & child health index	−0.84	−1.32	−0.49	−1.00
Part I crime rate	112.6	94.6	89.9	73.9	% of births where mother received prenatal care in 1st trimester	56.1	52.0	58.4	65.9
					% of births w/satisfactory birth weight	87.4	80.9	85.1	84.7
Violent crime rate	46.5		34.5	28.4					
Juvenile arrest rate	210.7	220.0	189.6						
Juvenile arrest rate: drug related offenses	68.0	99.7	69.4		% of births to term	80.8	82.1	82.0	83.0
Juvenile arrest rate: violent offenses	18.5	12.6	19.8						

Workforce and Economic Development (SI = Commerce)

	2001	2002	2003	
% commercial properties w/rehab investment of $5,000+	1.3	3.1	2.5	Additional indicators for *Workforce and Economic Development* can be found at the zip code level. Go to *Vital Signs 3* and download the *Workforce* section.
% vacancy among commercial properties	11.8	13.7	14.4	

Sanitation: Justice/Norms

	2002	2003		2002	2003
Rate of illegal dumping incidents	23.0	13.4	Rate of abandoned vehicle incidents	22.2	28.0
Rate of dirty streets & alleys incidents	21.4	42.2	Rate of rat incidents	12.8	20.2
Rate of clogged storm drain incidents	4.7	5.7			

(*continued*)

Table 10.6. (*Continued*)

Urban Environment and Transit: Flora		
		2001
Citywide indicators for *Urban Environment and Transit* can be found in the Baltimore City profile.	% tree canopy coverage	1.4

Education and Youth: Education/Age/Status

	2003 MSA School Test Scores				Attendance	2003 High School Achievement	
	Reading		Math		Absentee Rate		
	B	P/A	B	P/A		Grade 12 high school completion rate	83.3
Grade 3	50.7	49.4	50.0	50.0	21.79	Dropout rate	10.3
Grade 5	69.2	30.8	62.1	37.9	16.40	Advanced programs: University of Maryland	56.0
Grade 8	62.6	37.4	85.3	14.7	64.86	Advanced programs: tech/career	13.1
Grade 10	74.1	25.9	91.4	8.6	57.24	Advanced programs: both of the above	26.2
						Enrollment	
						Grades 1–5	827
						Grades 6–8	552
B=Basic P/A=Proficient/Advanced						Grades 9–12	551

Neighborhood Action and Sense of Community: Cult. Res.—Organizations

	2003		2000 General Election	2010 General Election
Neighborhood associations	16	% of population ages 18+ who registered to vote	55.1	78.3
CDCs	2	% of population ages 18+ who voted	27.7	30.6
Community gardens	1	% of population ages 18–25 who registered to vote	45.2	46.6
CHAP properties	0	% of population ages 18–25 who voted	16.6	12.5

Table 10.7. North Baltimore/Guilford/Homeland social profile (data from sources other than census)

Family Cycles	2000	2001	2002	2003	Children and Family Health, Safety, and Well-Being	2000	2001	2002	2003
Domestic violence rate	10.8	13.3	12.3	6.7	Teen birth rate	1.5	1.5	1.5	1.5
Child abuse rate	1.9	2.2	N/A	N/A	Maternal & child health index	1.18	0.78	0.61	0.58
Part I crime rate	57.2	48.1	50.8	29.5	% of births where mother received prenatal care in 1st trimester	94.2	90.2	93.5	96.2
Violent crime rate	7.1	5.7	10.8	4.1	% of births w/satisfactory birth weight	96.8	92.1	92.8	88.7
Juvenile arrest rate	41.0	35.0	40.0	31.0	% of births to term	91.6	90.9	91.5	89.2
Juvenile arrest rate: drug related offenses	9.0	6.0	13.0	6.0					
Juvenile arrest rate: violent offenses	4.0	3.0	4.0	4.0					

Workforce and Economic Development (SI = Commerce)

	2001	2002	2003	
% commercial properties w/rehab investment of $5,000+	4.4	1.8	0.9	
% vacancy among commercial properties	0.0	0.9	0.9	Additional indicators for *Workforce and Economic Development* can be found at the zip code level. Go to *Vital Signs 3* and download the *Workforce* section.

Sanitation: Justice/Norms

	2002	2003		2002	2003
Rate of illegal dumping incidents	1.6	0.8	Rate of abandoned vehicle incidents	6.2	9.6
Rate of dirty streets & alleys incidents	1.0	3.4	Rate of rat incidents	2.1	4.0
Rate of clogged storm drain incidents	1.8	6.8			

Urban Environment and Transit: Flora

	2001
% tree canopy coverage	37.3

Citywide indicators for *Urban Environment and Transit* can be found in the Baltimore City profile.

Education and Youth: Education/Age/Status

	2003 MSA School Test Scores				Absentee Rate	2003 High School Achievement	
	Reading		Math				
	B	P/A	B	P/A		Grade 12 high school completion rate	87.5
Grade 3	17.2	82.8	20.7	79.3	3.45		
Grade 5	20.8	79.2	29.2	70.8	7.41	Dropout rate	7.1
Grade 8	23.8	76.2	42.9	57.1	4.35	Advanced programs: University of Maryland	87.5
						Advanced programs: tech/career	4.2
						Advanced programs: both of the above	8.3

(*continued*)

Table 10.7. (Continued)

Grade 10	47.8	52.2	59.1	40.9	12.00		Enrollment
						Grades 1–5	141
						Grades 6–8	98
						Grades 9–12	116

B = Basic P/A = Proficient/Advanced

Neighborhood Action and Sense of Community (Cultural Res. = Organization)

	2003		2000 General Election	2002 General Election
Neighborhood Associations	24	% of population ages 18+ who registered to vote	53.7	54.4
CDCs	1	% of population ages 18+ who voted	41.0	39.3
Community Gardens	0	% of population ages 18–25 who registered to vote	14.1	13.4
CHAP Properties	3	% of population ages 18–25 who voted	6.5	5.0

Table 10.8. Baltimore City: social profile (data from sources other than census)

Family Cycles	2000	2001	2002	2003	Children and Family Health, Safety, and Well-Being				
					Health	2000	2001	2002	2003
Domestic violence rate	57.1	58.7	54.3	43.7	Teen birth rate	84.9	81.1	77.1	68.2
Child abuse rate	14.1	13.4	N/A	N/A	Maternal & child health index	0.00	0.00	0.00	0.00
Part I crime rate	106.0	100.1	88.1	79.4	% of births where mother received prenatal care in 1st trimester	72.1	72.1	76.5	74.9
Violent crime rate	26.2	23.7	22.1	19.8					
Juvenile arrest rate	127.5	136.9	145.6	140.1		86.2	86.5	86.5	86.3
Juvenile arrest rate: drug-related offenses	39.3	46.0	47.3	46.8	% of births w/ satisfactory birth weight	84.4	85.2	85.6	85.4
Juvenile arrest rate: violent offenses	10.7	11.2	11.7	9.6	% births to term				

Workforce and Economic Development (SI = Commerce)

	2001	2002	2003	
% commercial properties w/rehab investment of $5,000+	3.7	4.6	4.2	Additional indicators for *Workforce and Economic Development* can be found at the zip code level. Go to *Vital Signs 3* and download the *Workforce* section.
% vacancy among commercial properties	4.4	4.9	5.1	

(*continued*)

Table 10.8. (*Continued*)

Sanitation: Justice/Norms

	2002	2003		2002	2003
Rate of illegal dumping incidents	9.5	5.4	Rate of abandoned vehicle incidents	24.5	37.6
Rate of dirty streets & alleys incidents	6.1	17.2	Rate of rat incidents	7.7	14.9
Rate of clogged storm drain incidents	3.7	5.2			

Urban Environment and Transit

Air	2000	2001	2002	2003	Flora	2001
"Code Red" days	4	10	16	2	% tree canopy coverage	19.9
Days above 90 degrees	11	22	51	14		
Hazardous waste sites	103	112	120	115		

Education and Youth: Education/Age/Status

	2003 MSA School Test Scores				Absentee Rate	2003 High School Achievement	
	Reading		Math				
	B	P/A	B	P/A			
Grade 3	45.4	54.6	45.8	54.2	13.95	Grade 12 high school completion rate	81.4
Grade 5	50.1	49.9	56.2	43.8	13.18	Dropout rate	11.7
Grade 8	57.6	42.4	81.0	19.0	35.72	Advanced programs: University of Maryland	61.4
Grade 10	64.5	35.5	83.1	16.9	51.23	Advanced programs: tech/career	10.8
						Advanced programs: both of the above	23.3

B = Basic P/A = Proficient/Advanced

Enrollment

Grades 1–5	34,613
Grades 6–8	22,007
Grades 9–12	25,576

(*continued*)

Table 10.8. (*Continued*)

Neighborhood Action and Sense of Community (Cultural Res. = Organization)

	2003		2000 General Election	2002 General Election
Neighborhood associations	650	% of population ages 18+ who registered to vote	58.1	59.0
CDCs	45	% of population ages 18+ who voted	35.7	32.6
Umbrella organizations	54	% of population ages 18–25 who registered to vote	41.7	40.1
Community gardens	25			
Parks groups	35	% of population ages 18–25 who voted	17.1	12.9
Strategic nghbrhd action plans	6			
Healthy nghbrhd initiatives	10			
Main street initiatives	10			
CHAP Properties	7,259			

6

In 1996, an interdisciplinary group of researchers began to develop a research project to submit to the NSF for a new urban LTER program and an Environmental Protection Agency (EPA) research program on interdisciplinary studies of urban watershed systems. The effort involved scientists from the Carey Institute of Ecological Studies, Yale University, and the U.S. Forest Service. If the research grants were secured, then scientists from other universities and institutes would be involved. The Carey Institute would remain as the lead organizing and operational institution.

At the time this was an unprecedented venue for ecological research and such a diverse combination of researchers. There were many challenges to resolve before the proposal could even be designed. One colleague noted that the challenges made the tangled work of committees in UNESCO look like "a walk in the park." There were questions of theory, methods, sampling approaches, and concepts that needed to be reconciled within biophysical and social science disciplines as well as the big gaps between these two major groupings. A modified HEM model became the means for conceptual sharing within and between the social and biophysical disciplines.

Several of the team members noted that the work was one of the most frustrating and most exciting efforts in their intellectual career. The first submissions were not accepted. After revisions and resubmission, both the NSF and EPA proposals were accepted. The LTER has since been renewed three times. The human ecosystem framework continues as an integrator and interface for the NSF-LTER. The LTER directors note: "Our information and data management approach is centered around an adaptation of the Human Ecosystem Framework (by Machlis, Force, and Burch) called the Human Ecological System (HES). The HES ... interface serves as a structure by which we will integrate, or 'hang' Baltimore Ecosystem Study data in a meaningful way ... our hope is the HES will stand as our primary mechanism of data transfer thereby solidifying that notion that urban ecology requires a systems approach and when integrated these data will yield a more meaningful and accurate characterization of the Gwynns Falls Watershed and the ecology of the Baltimore Metropolitan Region."[9]

These hopes are a replay of the intent in the first publication that was produced by some of the LTER researchers who were using the framework in the design of their proposal. They say: "The framework consists of the human ecosystem model, which combines the ecosystem concept derived from biologically based ecology with insights about social institutions and interactions... the framework recognizes the openness of ecosystems, and does not assume equilibrium or self-regulation of urban ecosystems. The framework is completed by recognizing spatial heterogeneity in both the natural and social components of the urban ecosystem."[10] They go on to note that both systems are dynamic, can be hierarchically organized, and can use a watershed approach for assessing the interaction between social and biological components of an integrated system.

7

The urban habitat was once the prized place of political, spiritual, and economic power in the world. Only a minority of humanity existed in such venues throughout most of the history of our species. In the early decades of the twenty-first century, for the first time cities became the primary habitat for the majority of the world's human population. In economically emerging regions, cities are the most rapidly growing venues of human hopes and material desires. If this dramatic transformation to such a new preferred habitat had occurred for any other species, there would be massive amounts of research funds as well as studies and newly emergent biological specializations. Yet we ignore this dramatic transformation without considering the tremendous struggles that are occurring as adaptive strategies are tried out and fail and new ones emerge. Signals from overstressed emergent social and environmental urban places suggest a difficult transformation. Often, the pain is extended and increased as we continue to use the academic "egg carton" approach to policy and management where each discipline and administrative department resides in its own protected cell. Individual psychology is in this department, teaching and learning is in this one, family welfare in this one, and health and policing over there. However, reality demands an escape from this compartmentalized vision.

The reality is that a city is an ecosystem, and each element is shaped by the characteristics of the other. This change signals that all of the world is now in some way a part of the urban life plan. In the Baltimore case study, we have tried to anticipate a more organized way to develop research, policy, planning, and management concepts for a smoother and less painful development. We have tried to illustrate how the HEM can help us not simply to segment problems as if they had no relation to one another but to look at the entire ecosystem. We are suggesting techniques to combine elements and to link interactive effects of system elements, to include biophysical, infrastructural, and sociocultural elements in a more effective, efficient, and equitable manner. Our goal is to solve problems in terms of the actual reality we must work with rather than what we might hope to have. We are certain that the participatory approach of organizations such as the Baltimore Neighborhood Indicators Alliance is essential. Our interest is to join data that emerge from such activities into a dynamic model that permits decision makers and researchers to find the key points of "where to hit it" in terms of a functioning and sustainable improvement in quality of urban life. The HEM is a process, not a finite solution.

E·L·E·V·E·N

Extending the Capability of the Model

1

How can the HEM be made more useful? We might begin by building better capability in forecasting the likely consequences of natural resource and environmental decisions. Like most HEM work, this effort is part of a learning process. For example, what explanatory gains might be realized if we connect the structure of the HEM, the empirical reality of adaptive management, a tightly focused conceptual framework (patch analysis), and a basic theory (population dynamics)? These linkages are likely to improve our range of options and have greater long-term benefit and the least negative consequences. The goal here is not just to know what is connected but also to identify which connections are most critical in balancing risk and benefit. Given the complexity of human ecosystems, most of our decisions will remain within some level of uncertainty. However, if our exploration works, then our uncertainties can be reduced, and we will have a better idea of the long-term consequences. In this chapter we provide an exploratory test.

At the core of this analysis are the hierarchical variables of territory, status, and power, along with the identity variables of age, gender, and class, all of which come from the social order component. We connect these to the population variable from the socioeconomic resource component. The biophysical resources component provides a shifting set of variables such as land, energy, air, water, and flora and fauna, as do the individual and institutional cycles of change and the flows of individuals (immigration and emigration).

In this test case, we seek to develop a "logic of inquiry" that improves the predictive capability for future applications of the HEM. We emphasize that these are mutually shared questions and measures used by both social and biophysical scientists.

The exploration draws on insights from three established centers of thought, action, and explanation. The first is the *patch analysis framework*, which biologists use to understand stability and change in forest ecosystems and have applied to understanding urban ecosystems. The second is *adaptive management*, which is "based on a structured learning and adapting" approach in which uncertainty is a critical element in sustainably managing resource systems. The third combines *population theory*, *demographic measures*, and *territorial studies* to give us a better understanding of the mechanisms that regulate spatial behavior at the local scale.

These three approaches operate from a common ground similar to that which motivated the development of the HEM. All recognize co-variation between human systems and biophysical systems. They consider behavior at the local level as a critical factor in sustaining the system. All look at the empirical world as it is—an ever-moving, ever-changing set of interactions that do not move toward a fixed equilibrium "climax" state but rather follow a script of changing adaptive patterns. As William Cronon notes, "if all ecological change was either self-equilibrating (moving toward climax) or nonexistent (remaining in the static condition of climax), then history was more or less absent except in the very long time frame of climatic change or Darwinian evolution." The shift in perspective of the three approaches permits a better goodness of fit between human and biological histories. Ecosystem surprise and uncertainty are now part of the shared dialogue. The human story now becomes part of the biological story, as the biological story can be part of the human story. In this way we can develop an adaptive strategy to produce some level of goods, benefits, and services from the ecosystem; not through command and control but rather from positions as learner, interpreter, and facilitator of the adaptive process most likely to permit the system's continuity. This view of long-term sustainability happens only when sufficient human energy is put into learning and adapting.[1] For example, adaptive management includes humans (in decisions and learning) but has rarely taken on the fuller integration of ecological welfare (such as legal requirements concerning wildlife issues). The HEM may help to direct us toward that fuller integration.

2

Pickett and Mary L. Cadenasso define patch analysis as having two components: "First is the existence of patches—on land or in aquatic systems—that differ from one another in species composition, physical structure, or ecological processes. The second component of the definition is the fact that individual patches change through time, as a result of succession, or as a consequence of the movement of materials, energy, and organisms among them. Third if individual patches change, then so too will the entire array of patches."[2]

Pickett and Cadenasso stress that the utility of patch analysis for urban planners cannot occur without linking it to the context of the system—its "meaning" as an ecosystem that organizes a specific model of how the system works and the metaphor of the "mosaic" of the larger system within which the patch exists.[3] Much of their attention is directed to the "flux" of materials, energy, genetic information, and so on within the defined patch and between elements of the larger mosaic. In our earlier chapters, we have examined social patches in a manner similar and with similar concepts to those used by Pickett and Cadenasso. As we noted in chapter 6, the demographic flux of the post–World War II period caused by the "boomer" generation created a well-tracked flux affecting most social institutions in the United States. One could follow this "pig in the python" as demand on various institutions is swallowed and then passed on by the boomer cohort to the next phase. As demand moved on to other needs, schools went from too short in supply to too plentiful. Today the political debate is responding to the arrival of this generation at retirement age with increased demand on health services, hospices, and so on, with the morticians awaiting their boom.

Later in Pickett and Cadenasso's report, they emphasize the importance of models for applying a framework like patch analysis. They identify two models: One is a modified version of the HEM, and the other is a patch mosaic classification of a neighborhood landscape in Baltimore. Though their use of both models clarifies some connections between social and biophysical elements of human ecosystems, there is only a limited means for identifying the reciprocity within and between these elements of the urban ecosystem. They make no direct connection between the spatial patterns they observe and the underlying dimensions and drivers of the human elements they map. So our first task will be to add some empirical depth to those dimensions and to use the patch framework to ground the abstractions of the HEM.

Adaptive management is the next strategy to define. We have some idea about how human ecosystems change through internal and external perturbations and how biophysical and sociocultural elements are intertwined participants in these changes. The question remains how to aid human ecosystem response to move in a more sustainable way. Adaptive management is a strategy that has been practiced for over a quarter-century. Its lessons seem most fruitful for application of the HEM, patch analysis, and causal premises of demographic research. Where it works to improve long-term sustainability of the human ecosystem, adaptive management becomes one of the most important ecosystem processes.

Bernard Bormann and colleagues argue that since the introduction of the concept of adaptive management, it has been defined in many ways and has different meanings for different people. Recently there appears to be a consensus on its definition and recognition of it as a scientific discipline using the best available science when it is applied.[4]

Adaptive management is a decision framework where uncertainty is recognized, knowledge is gained through science-based, structured learning that is integral to the management action and where questions linked to potential future decisions drive monitoring to increase the chances that what was learned will be widely adopted. Involving stakeholders as co-learners is so important that collaborative adaptive management is generally seen as the way forward.[5]

Demography is the study of population changes at the local (patches) and global scales. It examines how fluctuations in birth and death rates and emigration and immigration rates alter the rules, regulations, and relationships of a given social unit. It permits the development of census data that can generate predictive indicators such as birth and death rates sorted by certain social characteristics. It collects quality of life indicators such as dependency ratios, housing conditions, income and status, and access to health and educational services. For a biological perspective, one can gain insight from chapter 2 in G. Evelyn Hutchinson's *Introduction to Population Ecology*. The careful reader can see how this pioneering ecologist moved between social and biological data to understand basic demographic processes. All social systems need knowledge on the patterns and processes affecting population characteristics so that they can anticipate policies and practices that need to emerge for response to short- and long-term effects of specific population changes. So data are universally available, though they differ in quality. The modern technologies of information collection and gathering permit great analytic and adaptive capability.[6] We will consider two dimensions of demographic influence—territory and social solidarity.

3

There have been many disasters of hurricanes, floods, heat waves, wars, and revolutions impacting cities of the twentieth and early twenty-first centuries, and yet the astounding consequence is that out of massive destruction of persons, infrastructure, and ways of movement, the residue of family structure, institutional practices, and basic solidarity at the patch scale emerge to re-create urban systems.[7] Consider the resilience we observe in city neighborhoods in responding to the intensive bombing raids over London, Dresden, Tokyo, or Fallujah. This resilience seems built on the underlying social structures that have been fulfilling the daily, mundane social needs of a given social patch system. In crises such as natural disasters as well as wars and other human-made disasters, people regroup, reorganize, reestablish a normative structure, and get on with helping one another. Although the social patches that retain resilience tweak our interest and admiration, most often we devote our research to recording those social groupings that fail to recoup. The real Darwinian evolutionary triumph is this cooperative natural survival pattern provided by elements of social patches rather than the social Darwinism response we often expect. The response patterns and processes here are similar to those observed in forests responding to patch-scale fire or other perturbations where enough genetic memory permits rehabilitation.

In both biophysical and social patch analysis, we examine the variables that sustain one patch in the face of great stress. We also try to determine what the missing factors are in the case of similar patches facing similar stress, but fail the resilience test and do not recover. What holds a human ecosystem together so that it avoids tipping toward failure will be one of the central questions for future application of the HEM. In discussing biophysical patches, we interpret this as the ability of the elements of the eco-units to be aware of one another and to have certain expectations of the behavior of the other elements.[8] That is, there is a division of labor that the system needs to sustain "trust," and this permits a "natural" order out of seemingly disparate elements that individually may vary greatly, but, in combination, they form a distinct, resilient, and ongoing pattern whose regularized processes reinforce stability.

Most studies of ecosystem behavior are seen as being bound by the laws of physics and the rules of Darwinian evolution rather than some idea of "free will" guiding a routinized mix of irrational and rational decisions. This attitude is characteristic of many archaeological, architectural, historical, and sociological studies of cities over a wide array of cultures, times, and places. These studies suggest a universal consistency in the clustering of ecosystems in terms of their functional and symbolic aspects. This spatial clustering of commercial and institutional activities, ethnic, and social status residential areas assumes basic benefits and efficiencies or the need to be together for mutual support as "natural" relatively homogeneous social units. As Timms says, "The urban community is neither an undifferentiated mass nor a haphazard collection of buildings and people. In residential differentiation of the city the urban fabric comes to resemble a 'mosaic of social worlds.'"[9]

This clustering of activities and social groups, this "patchiness" seems to be a universal adaptive strategy for social species of plants and animals. In one sense, we continue our hunter-gatherer tradition of first responding to the norms of our primary small-scale community within the larger mosaic of social opportunity. People in one patch would migrate in a stepping stone fashion where early migrants from a given clan or village would establish a beachhead, and then other members would gravitate to that place.

Ecology as a discipline was being created and nurtured by both biologists and sociologists in the early twentieth century. Both disciplines were looking at whole systems composed of concentrated patches of behavior. The University of Chicago human ecologists identified "natural areas" in downtown Chicago that were based on clustered types of economic activity, social class, and ethnic cultural clusters. Concepts such as climax, succession, and displacement were metaphors borrowed from biology to account for the observed cycles of change, decline, and renewal in central cities. This borrowing and sharing of concepts was a means for treating the observed spatial patterns as determined by forces outside the realm of rationality and emotion. It was more "science-like" to treat these as givens. Many biologists have been comfortable with this frame of social science where familiar concepts are used and much of the observed human behavior is predetermined.

However, other scholars responded by noting that the predictability was in government rules and the underlying emotions of racism and bigotry; thus, the prevailing schemas missed much of the lifeblood that drove human creativity and adaptability and made emotion and irrationality a necessary part of human choice. David Ley summarizes the evolution of the first two phases of human ecology "theory" and research approaches. Though Ley acknowledges the depth of empirical work by the early human ecologists, he suggests that the "natural area" approach had shortcomings, as "they were heavily biased toward the central sections of the city where unambiguous lifestyle and ethnic territories were contained . . . [but] . . . not directed to less locality-based groups or to suburban districts, where less precise spatial articulation checked the easy demarcation of natural areas." Later he notes that the use by the Chicago School of human ecologists of only a few indicators led to "loss of the distinctively human character of place and a tendency to generate typologies which, though suited to a specific academic purpose, created regions that were often less recognizable as lived places."[10]

Ley traces the intellectual trends to refine spatial analyses through Burgess's concentric zone model, Hoyt's sectoral model, the social area model of Shevky and Bell, factorial ecology studies, and work by Firey and others on the "symbolic" nature of observed urban landscape clusters and their role in shaping and sustaining the larger system. Ley notes: "Inherent in ecological models is a somewhat mechanical and deterministic view of the development and evolution of social areas according to market processes that neglects, among other things, the cultural and volitional aspects of urban life."[11]

Michelson in an early review of studies on social behavior and the urban environment notes "that physical variables and their interrelation with social variables have been largely neglected by human ecologists." He goes on to observe: "Space has been utilized as a medium in most of human ecology rather than as a variable with potential effect of its own."[12] Environment then is confined to social aspects rather than biological or physical. In a classic work on ecological history, Cronon argues:

> If we avoid assumptions about environmental equilibrium, the instability of human relations with the environment can be used to explain both cultural and ecological transformations. An ecological history begins by assuming a dynamic and changing relationship between environment and culture, one as apt to produce contradictions as continuities. Moreover, it assumes that the interactions of the two are dialectical. Environment may initially shape the range of choices available to a people at a given moment, but then culture reshapes environment in responding to those choices. The reshaped environment presents a new set of possibilities for cultural reproduction, setting up a new cycle of mutual determination. Changes in the way people create and re-create their livelihood must be analyzed in terms of

changes not only in their social relations but in their ecological as well.[13]

At the patch scale of organization, biologists tend to see the patch as a means for sustaining biodiversity. Perturbations like fire or insect invasion release plants and genetic factors that keep the diversity of the system. Social scientists might see the patch as being homogeneous at the local level and diverse at the level of the district or wider social unit. The shared goal is about encouraging and sustaining diversity. Biologists want to stir up things in the ecosystem to create and sustain biodiversity at the larger scale, while the social scientist is concerned about encouraging cultural diversity. The adaptive capability at the higher scale is being stimulated to adopt new patterns or to fit older patterns into newer challenges. It is this unity of goals that should drive our mutual search for understanding the human ecosystem.

The challenge for human ecosystem applications is to be aware of these regularities in clustering and their impact on human behavior and the biophysical environment. There are certain genetic predispositions. There are choices within a range of "must do," "must not do," and "maybe." There are norms that direct human behavior, and similar flexibility is found in other social species. It is that range that may encourage creative rearrangement of the life chances given by genes and other stern determinants. It is in these creative openings where some pioneers get to reshuffle the cards.

4

The demographic "flux" of human populations can be rigorously measured and observed in a variety of human ecosystems. See, for example, William Kornblum's study of blue-collar communities in South Chicago. He analyzes the shifting ethnic mixes over time in seven South Chicago neighborhood patches. He measures the flux from predominately Irish and Polish immigrants in 1900 to Polish in 1930 and Mexican, black, and Polish in 1970. His analysis is similar to the work by many sociological studies on the flux of urban immigration and migration.[14] Pickett and

Cadenasso have a similar analytic approach of succession in their studies of biological patches in Baltimore. Earlier we saw how Klinenberg used a framework similar to Kornblum's in examining similar data on varied responses to the Chicago heat wave.[15]

One of the interesting findings from a variety of studies is the stability of patches where some ethnic groups moved into the neighborhoods as others moved to outer neighborhoods. Firey also found similar trends in Boston where there were shifting patterns of stability with change in the ethnic composition, where the traces of prior occupant norms remained and many neighborhoods sustained its prime identity. There was flux, but it was channeled into familiar normative and social status levels.

Kornblum documents the succession by newer ethnic groups into many of the social patches comprising South Chicago. He identifies the social mosaic of a region in transition: "In general, the entire South Chicago area is honeycombed by neighborhoods which differ according to ethnicity. Included among its residents are Serbians, Croatians, Poles, Italians, Scandinavians, Germans, Mexicans, and blacks, who make up the main residential groups . . . there is further segregation on the basis of generation of arrival in the city."[16]

Kornblum finds that this spatial mix of ethnic groups is shaped by the political institutions laid down by the early arrival generations: "In becoming involved in political contests generations of South Chicago people have continually modified their definitions of who 'belongs' in the community. And through the negotiation of new primary groups in local politics South Chicago people also create a blue collar culture which all local groups, even those who are initially the most feared, eventually come to share."[17]

In sum, there are institutional mechanisms for managing flux in many of the basic social ecology components of the larger regional mosaic. While the ethnic groups sustain their identity with their "traditional" religion, food, music, and language, the political mechanism steers exclusion into inclusion. This absorption into some traces of norms and actions created by the prior occupants of their social patches is similar to those "historical" traces found in biological systems.

Kornblum notes the positive dimensions of the waves of ethnic migrants living in neighborhoods and seeking well-paying jobs for persons without high levels of educational attainment. However, these benefits seem at risk when the government uses certain criteria to evaluate the locations for wise and poor investment in housing. Douglas Rae's thoughtful analysis of patterns of "race, place, and spatial hierarchy" in New Haven documents how government zoning to help sustain middle-income owners of single family residences took the informal spatial patterns and converted them to fixed and certified formal patterns.[18] The mortgage certification areas combined with the loss of factories and the great migration of blacks cancelled the transitional neighborhood pattern observed by Kornblum. The new arrivals in New Haven found segregated places with declining occupational opportunity and public housing without formal policing of the hallways or the informal social control of the neighborhoods. He argues for reinventing these vernacular networks.

Mark Abrahamson's study of urban enclaves gives more details on how the institutions in Boston's Beacon Hill sustained the status of that social patch. In the same work, he gives a good analysis as to how "urban renewal" policies destroyed community social institutions and forced the residents into architectural deserts with no ability to create the older traditions. In a more detailed study, Harold A. McDougall noted: "Blacks in Baltimore remember segregation as a cruel symbol of inequality, but they also remember it as a context for stable black vernacular neighborhoods, with a strong work ethic and closely networked local economies."[19]

We note the similarities in the biological and sociological approaches to social patch behavior. At the scale of patches—social and biological—we can ground the HEM with familiar response patterns. The demographic flux and response in the social patch represents resource flows that are measured by the rise and fall of certain kinds of jobs, materials, and energy flows. The institutional structure canalizes the patch response to flows into and out of particular patches to sustain the continuity of the patch (neighborhood). In both cases, there is a measureable reciprocity in the biophysical elements of air quality, land use, and infrastructure capability. In short, there is predictable covariation between the social and biophysical elements of these patches in response to specific trends in population flux. Here it is clear that the social and biophysical researchers are "sharing" similar perspectives about the determinants of ecosystem pattern and process. They are entwined, not borrowing one another's concepts and approaches.

Klinenberg's "social autopsy" of the 1995 heat wave disaster, discussed in detail in chapter 7, is a good complement to Kornblum's study. Klinenberg gets beneath city-wide statistics to tease out the value of locally based patch structures and the consequence of variations in neighborhood and place ecologies. He treats the larger urban scale as context within which he looks at smaller units such as neighborhoods and patterns within these neighborhoods to get at cause. As discussed in chapter 7, his analytic focus is sharpened by an in-depth look at two adjacent residential areas with similar social characteristics but dramatic differences in death rates. It is important to note the valuable empirical and theoretical understanding that Klinenberg's work is providing for HEM application and grounding by patch analyses.[20]

In the Chicago neighborhood of Little Village, the people moving out were quickly replaced by new but somewhat similar immigrants, in a pattern similar to those found in Kornblum's 1970s research. The result in terms of human health was significant when compared to North Lawndale right next door. Here the urban ecology was one of weed-filled lots, empty structures, and widespread fear. In North Lawndale, the seniors were confined in poorly ventilated rooms and were fearful of venturing out, while in Little Village the active street life, including functioning public and private places such as shops and super markets with air conditioning made for public spaces that the elderly did not need to fear. As in the low-violence communities studied by Robert Sampson and his colleagues, the linkage of trust and busy street life sustained a sense of mutual protection.[21] This was simply not the case in North Lawndale. In the sub-neighborhood patches, the mutual support and awareness reinforced on a daily basis created "trust" in the safety of the busy streets and made one community far less dangerous and more supportive in the heat wave crisis than a similar community without these social capital elements.

The Chicago heat wave social autopsy demonstrates several variables that affect health and human welfare in communities and that reinforce the practical and scientific utility of the social patch analysis approach. Klinenberg states:

> Previous studies of heat wave mortality have shown that residents of places with high poverty, concentrated elderly populations, poor housing, and low vegetation are especially vulnerable ... several place-specific risk factors ... such as quality of public spaces, the vigor of street-level commercial activity, the centralization of support networks and institutions, concern the social morphology of regions; others, such as the loss of residents and the prevalence of seniors living alone, concern population-level conditions ... the key reason that African Americans had the highest death rates ... is that they are the only group in the city segregated and ghettoized in community areas with high levels of abandoned housing stock, empty lots, depleted commercial infrastructure, population decline, degraded sidewalks, parks and streets, and impoverished institutions. Violent crime and active street-level drug markets, which are facilitated by these ecological conditions, exacerbate the difficulties of using public space and organizing effective support networks in such areas.[22]

The work of Klinenberg and Kornblum are examples where strong empirical guidance can fill out human social patches with demographic and observational details on differences in emigration and immigration flows, institutional opportunities and constraints, and forces for integration of decline of these communities. We can use demographic techniques such as population pyramids where age classes for male and female members of the community over time give a strong means for prediction about the future need for various services and interventions. Indicator measures are abundant where sex ratios, dependency ratios, infant mortality rates, and similar measures define present challenges and future patterns. Many of these possibilities were demonstrated in chapter 9 in the discussion about the monitoring benefit of social indicators.

The familiarity of place interacts with and reinforces social institutions at the patch scale. When that connection remains adaptive or is destroyed at the patch level, it restructures the larger urban mosaic in predictable ways. Our understanding of this biosocial variation begins with the HEM menu that suggests some of the basic elements of the human ecosystem pattern and process that fit a given case. We then ground the HEM through application of the patch analysis framework using empirical work on population flux in urban social patches. Our next stage is building a theoretical system that helps to connect the parts.

5

The work of Émile Durkheim will serve as one example of how a social theory can be modified to explain regularities in the observed patterns of social organization. Durkheim, a late-nineteenth-century social theorist, suggested that flux from increased population and density would require institutional change to sustain a way of life. He posited two strategies of solidarity, or factors that hold together social patches: those of likeness or similarity of persons (mechanical integration); and those where differences and a division of labor compelled a more "organic" system of integration. The system is driven by its own necessity, as Durkheim notes: "The division of labor varies with the volume and the density of societies, and if it continues to advance in the course of social development it is because societies regularly become more dense and generally larger...We are not saying that growth and the increasing density of societies 'allow,' but that they 'necessitate' a greater division of labor. They are not, for the latter, an instrument of its realization, but rather the determinant cause of it."[23]

In short, a shift in population size and density within a given territory compels an ever-increasing division of labor to sustain the system, while at the local level the sustaining of a patch identity requires a measure of mechanical solidarity. This seems the point where the methods of meaning, model, and metaphor discussed by Pickett and Cadenasso join their work to social ecology.

Durkheim's theory, when modified to fit the realities of a twenty-first-century global worldview rather than the more segmented world of nineteenth-century rural and urban life in the nineteenth century, gives us a means for using computer-driven demographic information to predict likely future directions of a given social patch. As population density (persons per square kilometer) varies from high to low, new organizational forms must emerge to sustain the human ecosystem. However, this pattern in the larger system is also played out at the neighborhood or village level. As elders or ethnic groups are clustered in a geographic patch at a certain density, the area's continuity is sustained by the specialized shops, cafes, schools, religious centers, and other social mixing and sustaining venues. As populations migrate in and out of these neighborhoods, territorial expansion or reduction becomes a predictable pattern. The new infusion or loss of former residents shifts the social patch with new vigor or changes it into a territory where mechanical solidarity is lost to the caretaking of organic structural elements from public agencies distant from the social patch.

A modification of Durkheim's theory would suggest that the downshift in population size, density, and diversity of neighborhoods greatly reduces the potential for mechanical solidarity to continue to buffer the context of changes in the organic solidarity of the larger urban system. In this case the representatives of the contextual organic solidarity remain the primary operational unit—first responders, police, social workers, health emergency workers—in short, social service workers. Hence, in Chicago's North Lawndale, the dissolved institutional structure was not available to sustain the trust necessary for safe and functioning public spaces—parks, sidewalks, shops, and other meeting nodes. It is important to note that for the city at large the highest death rates were in predominately African-American areas, but those African-American neighborhoods that had low population loss—Riverdale, Auburn, Gresham, and Calumet Heights—saw only modest death rates. In short, it is not simply a racial factor but a strong association with social ecological characteristics.

Steve Johnson's essay clearly notes the substantial and important influence on his theory of the street-level insights of Jane Jacobs, where the choreography of city streets establishes these patterns or rules of behavioral interaction between strangers.[24] The common recognition of shared space permits people to follow daily routines with a high degree of expectation that things will work out as anticipated. The wonder is that most often, they do. We only become aware of these patterns when they are violated by the ignorant or the arrogant. Johnson stretches his theory to include other naturalistic social behavioral connections such as ants, slime mold, and computer software programs, and in all of these systems there is the importance of "neighbors" at the local scale who sustain the familiarity of belonging to the neighborhood and whose patterns follow simple emergent, informal rules, and at the higher-scale level illustrates that local self-organization patterns extend upward as well as trickling down.

Neighborhood patches are held together by their members' sense of comfort, familiarity, and general trust in one's ability to predict the likely behavioral outcome of routine social transactions. Trust comes from simple, routine solutions to daily life where one believes the other person will pass on the correct side, that you recognize that stranger because they regularly recognize you in commercial transactions or from the simple act of passing one another on the street with a wave or smile. These are Jacob's ideas, that interesting edge effects generate busy traffic, frequent interaction, and therefore many eyes on the street. These are the base for the links in neighborhood efficacy. Sampson and his colleagues demonstrate how local social control manages a big issue like street violence.[25] However, the resolution of that issue depends on the many simple and modest actions that establish the trust between strangers observed by Jacobs and Johnson.

There are many variations on how mechanical solidarity based on likeness serves important functions for certain neighborhoods. There are many patches where social class is a critical determinant of who lives in such an area. These are areas where external regulation of behavior is mostly enforced by police, security guards, and other bureaucratic elements. In these situations, an informal awareness and regulation of the dance of strangers is not the governing force; indeed, the hint of diversity and difference becomes an alarm upon the entrance of a stranger whose difference calls for the authorities to rectify the intrusion. Gated communities take it one step further, with physical barriers and elaborate internal security systems to ensure that only those who fit the rules of likeness are admitted. Durkheim's idea of "mechanical solidarity" becomes one of fragmented clans and tribes sustained by hired guns rather than informal norms.

6

There are many social patches where the homogeneity of impoverishment spatially clusters populations with characteristically high unemployment rates, poverty rates, percentage of female headed households, teen pregnancy rates, and other indicators of limited social resources. In poor countries like Bangladesh or the Philippines, one will find spatial clusters of much more materially poor urban populations with, however, an intact social institutional structure. In the United States, many of these clusters not only lack economic resources but lack functioning social institutions and other informal structures of support. Colin McCord and Harold Freeman report that death rates for males between the ages of 5 and 65 are lower in Bangladesh than they are in Harlem. Arline Geronimus and colleagues found that African-Americans living in some sections of Brooklyn and Queens, just a few miles from Harlem, had mortality rates similar to the average for white U.S. citizens. In comparison to the Brooklyn and Queens locales, the dangerous patches of Harlem are closer to aggregates of persons than communities, and persons or households must work out their own forms of adaptive survival and external control. In many such localities, the only viable economic activity available is trade in controlled substances, and the related turf wars are settled by gang struggles and mob enforcement. Sherry Olson's study of the making of Baltimore demonstrates that these sorts of residential areas have been a part of each historical economic wave.[26]

The ideas, theories, and findings of our social ecology authors gives us a series of *ex post facto* experiments that expand the utility of patch analysis. For example, what structures the emergence of social patches? What keeps them bound in space and time? And what permits the complex diversity of the entire urban system to sustain reasonable stability? As Poggi notes, "in spite of his frequent use and occasional abuse of biological analogies, Durkheim was not willing to entrust the realization of the new form of integration—'organic solidarity' . . . to the utter spontaneity of the evolutionary process. Hence his emphasis on the necessity of express, self-conscious, authoritatively elaborated, and sanctioned mechanisms of regulation."[27] That is, individuals are making choices that add up to the collective identity that depends on the pattern. The organization of those choices builds the larger integrative unit. In this broadened approach Durkheim is closer to Johnson and Jacobs than the organicist theorists.

Durkheim's notion of mechanical solidarity, where integration between strangers depends on likeness and exists not only in some gated housing complex or upper-class ghetto, but also in the buildings and blocks of the neighborhoods and rural villages in which they fit into the larger system's division of labor and resulting organic solidarity. Further, there is a better fit between the deeply emotional basis of social bonds found in the intimate human associations such as family and friends if they have reinforcing patch-scale social buffers between them and the larger urban ecosystem.

We are suggesting that there is a great variety of adaptive strategies within the small component units that make up a neighborhood or rural village within an urban or regional ecosystem. Some are based on likeness, some on emotional bonds, and some on the genetic and commercial ties of an extended family; these may form a sub-cluster, a tributary of the larger stream. With these bonds, seniors who monitor sidewalk use, mothers who together monitor the safety of children at the playground, men from nearby residences who look after the street trees or community gardens, and recent immigrants who open and close small shops give critical depth and dimension to the local social ecology. And that local reality is connected to a global reality that sustains local ties.

The work of Weiss and others have made use of fine-grain data sets at the level of zip codes—patch scales smaller than census tracts and that have more coherence of lifestyle. Grove and colleagues, noting the utility of Weiss' clustering technique, write: "This can improve the characterization of social groups, building from indices of population density, the ethnicity, or socioeconomic status to more complex characterizations." For instance, the PRIZM (Potential Rating Index for Zip code

Markets) categorization system created by a company called Claritias uses factor analysis and U.S. Census data about housing, household education, income, occupation, race/ancestry, and family composition to classify urban, suburban, and rural neighborhoods into categorical group measures.[28] Hence, there are statistical means for identifying these areas and measuring the lifestyle characteristics between them. By modifying Durkheim's theory of variation in social solidarity, we improve its predictive capability—as density varies, the nature of solidarity varies in a systematic way.

7

So far we have seen how the HEM variable of population can gain predictive capability by being applied at the small-scale social patch level using new statistical technology and as a part of a large-scale social theory on solidarity. Now we will consider another "driver" variable of the HEM—territory. Its relevance is particularly helpful as we modify the application of Durkheim's theory. The role of territory and spatial ordering of social behavior is well summarized in Carl Degler's analysis of the cycles of demise and revival for Darwinian ideas in social science studies. Most of his work focuses on studies from the 1960s and 1970s. In this period there was a strong interest in cross-species comparative studies of social and individual behaviors, in particular their relevance for policy decisions. Some social scientists worked with these ideas, and others resisted. Degler gives us an idea of the mix in acceptance and doubt about causality as he summarizes a talk by the sociologist John Doby: "There is an evolutionary relation between animals and human beings; it is true we have a 'genetic structure blueprint' just as all species do. Human beings, it is true, have a quite different structure from other species but it still is 'a mixture of mechanisms, some biological, some cultural' . . . and some not."[29]

His approach is very much the one we wish to follow. We do not want to tangle with the issues of nature versus nurture. We want to gain an understanding of the basic mutual dependence and reciprocity between people and their biophysical selves.

Territory often has much to do with matters of identity and the expansion and defense of boundary of the territory. Here we see how species and landscape variables can serve as referents of "appropriate" behavior and as indicators of the social status and ethnic identity of social patches.

Cross-species comparisons help us to understand certain universal behavioral patterns and organizational characteristics of social species such as ants, birds, large nonhuman primates, and humans and how their spatial patterns shape their survival strategies. Sociobiology may overstate the influence of genetic determinants on human behavior; however, its basic findings remind us that much of social behavior by all social species is routine and regularized through genetic or ritual mechanisms. Human ecosystems have highly predictable routines of rush hours and slack hours, school and sports seasons, and so on. Though these ritualized regularities may be determined by work/nonwork cycles, their patterns are not unlike those observed in ants and bees as they transport food or direct the members of their collective to new locations for hives and anthills. This predictability and response to barriers of travel are like those observed by news helicopters following the impact of an auto accident on a busy artery near a large urban hub late on a Friday afternoon.

As we have discussed, foresters, botanists, and other ecologists have demonstrated that the sustaining of the larger ecosystem such as a watershed is dependent on regular perturbation—such as fires, disease, and similar episodes—at the patch scale. In the forest examples, the stability and variability of the larger system require such renewal to sustain the overall genetic variability. The question here is similar to that asked by many human ecologists—what is it about spatial patterns that organizes and holds a human ecosystem together or leads it to its demise?

The value of cross-species comparison is evident in Wilson's early discussion of social spacing and territory. As he suggests, "Aristotle and Pliny noted the demarcation and defense of territories by male birds and the phenomenon was then sporadically rediscovered through the first centuries of modern science."[30]

Thus, our interest in patches and clustering of behavioral events has a strong historical base. Of particular interest for this book is his examination of the adaptive value of the "dear enemy" phenomenon, which suggests there are three mechanisms that make the adaptation possible.

The first is simply habituation, where other animals become aware of others by their regular presence. Wilson calls it a "form of learning ... in an almost literal sense, the neighbors 'tame' one another." The second mechanism he identifies is familiarity: "The probable adaptiveness of the effect can be easily guessed. The more something stays around without causing harm, the more likely it is to be part of the favorable environment." The third mechanism is convergence in dialect; for birds this expresses itself in song, and for people it is the accent in the local language. Misia Landau has argued that all scientific theories "arise out of a grand form of narrative [that], seek to reconstruct histories of a single entity—the cosmos, the planet, a species."[31] We see this in Wilson's narrative, where familiar neighbors train one another, and in some of the ideas expressed in the work of Jacobs and others as they look at what holds neighborhoods together with focus on the role of familiar strangers. Hence, our interest is in the power of familiarity.

Familiarity is a "natural" social experience; indeed, one might suggest that it is "hard-wired" as the great majority of humans who have ever lived have done so within the small social units of clan, tribe, or village. Until the early twenty-first century, the urban locale was a habitat for only a minority of the humans who were then living. Though the majority of humans now live in urbanized habitats the clan and village normative patterns still hold direction of behavior for most of us. For a classic example, see Herbert Gans's study of the West End "urban villagers" in Boston.[32] These are not actions of individual "free" choice, they derive from a tribal vision that has a much longer and more widespread history than urban market economies (a few hundred years versus some two hundred thousand years for "modern" humans).

The work of Kornblum and Klinenberg is enhanced by the studies on territoriality and personal space done in the 1970s. These studies had a widely accepted notion that tendencies toward violence were an innate characteristic for all species in expanding and protecting claimed territory. The rejection of innate violence and genetics as prime determinates of human behavior did not mean that social scientists did not do research on territorial behavior. They extended the riff on the territorial theme from the individual to larger social systems. On the sociological side Goffman noted, "Territories vary in terms of their organization. Some are 'fixed'; they are staked out geographically and attached to one claimant, his claim being supported often by the law and its courts. Fields, yards, and houses are examples. Some are 'situational'; they are part of the fixed equipment in the setting (whether publicly or privately owned), but are made available to the populace in the form of claimed goods while-in-use. Temporary tenancy is perceived to be involved, measured in seconds, minutes, or hours, informally exerted, raising questions as to when the claim begins and when it terminates. Park benches and restaurant tables are examples. Finally, there are 'egocentric preserves which move around with the claimant, he being in the center. They are typically (but not necessarily) claimed long term."[33]

In this latter case we are talking about personal space and its ability to regulate spatial behavior. Goffman's work used this frame to analyze points of conflict and unity in a wide range of human situations—parks, hospitals, prisons, and so on. E. T. Hall developed the theory of proxemics to examine how the human dance through social space is regulated. Both Goffman and Hall help to fit the individual into particular social patches. Anna Bramwell offers a useful context for understanding the ecological contributions to cross-species comparison as it moved science out of the lab and into the field.[34] Cross-species comparisons enable social scientists to move between the empirical concepts and observations made by nonhuman ecologists and those of anthropologists and sociologists such as Erving Goffman where people followed "appropriate" spatial closeness for a given situation. The genetic ties set the stage, and as Goffman notes, the observed clustering of behavior reflects adjustment to the norms of the public order. So in this sense, certain observed patch boundaries and expected behaviors are part of genetically shaped closeness that is both defining and reaffirming of patch structure and dynamics for social species.

Perhaps the best treatment of territoriality and how it structures social behavior is Torsten Malmberg's book *Human Territoriality*. It summarizes a wide range of studies that mix biological and social variables and crosses discipline boundaries and urban and non-urban venues. He provides an excellent summary of empirical findings from a variety of studies on the effects of spatial factors on the behavior of social life. He stresses the complexity of all such studies, noting that "the total environmental process can and should be studied from different vantage points and at different levels of analysis. This means, of course, that only a multidisciplinary approach can lead to a viable theory."[35]

The examination of research and theories from studies on territoriality among humans and other animals is important to the HEM analysis because the prime functions of patches are to protect spatial identity and to ensure comfort and security for the religion, music, cuisine, and language of primary reference groups. It is the combined strength of these social patches that shape the stability and change in the organization of the total social mosaic. This pattern is similar to the patch functions where small scale perturbations sustain the overall genetic diversity and stability of the overall forested ecosystem.

In a study on Chicago, Robert Sampson, Robert Raudenbush, and Felton Earls compared similar neighborhoods that had different rates of interpersonal violence. In looking at the neighborhoods that had less violence, they noted: "In contrast to formally or externally induced actions (for example police crackdown), we focus on the effectiveness of informal mechanisms by which the residents themselves achieve public order. Examples of informal social control include the monitoring of spontaneous play groups among children, a willingness to intervene to prevent acts such as truancy and street corner 'hanging' by teenage peer groups, and the confrontation of persons who are exploiting or disturbing public space ... it is [neighborhood] linkage of mutual trust and the willingness to intervene for the common good that defines the neighborhood context of collective efficacy."[36]

Though the system is composed of individual actions and interactions, it also has characteristics that transcend the individual patterns and process, an insight that is similar to the treatment of ecosystems by biological ecologists.[37]

Similarly, the Baltimore Ecosystem Study has done research that includes trees, woodlands, forests, and other vegetation as functioning members of the social community. They are favored members of the "social discourse." Trees and vegetation in cities reflect many social uses as well as their biological functions. Plant types and their arrangements have high utility as status indicators as well as historic, ethnic, or other symbolic links. As Gordon Whitney and S. D. Adams note, "Taste, fashion, and the relative availability of various species or cultivars are the major factors influencing the composition of the urban forest." W. Lloyd Warner was one of the earliest investigators to comment on the symbolic significance of the front lawn, particularly as it served as a guide to the taste and the status of the inhabitants. The close association of taste and status was later amplified by Bill Darrenbacher's study of residential vegetation in the San Francisco Bay region.[38]

As noted earlier, the more subtle dimensions of lifestyle clusters may come from market research data that identify neighborhood clusters of lifestyle consumer patterns and permits comparison between patches of consumerism. The marketing studies done on neighborhood lifestyle clusters is one means for understanding variation in support of environmental quality issues. Such studies reflect an expansion of the theory and methods used in understanding the role of the reference groups that guide our behavior.[39]

8

We have suggested that some of Durkheim's nineteenth-century theoretical ideas on urban density, division of labor, and "naturalistic" integration have a general goodness of fit with Jacobs's twentieth-century interest in neighborhood functioning; he even seems to have anticipated a twenty-first-century interest in the emergence of naturalistic clustering in "complexity theory." We have suggested that the examination of patch pattern and process in biophysical and sociocultural domains may help us to understand regularities in human ecosystems. We can use symbolic factors such as trees and other vegetation or traditional names of a place as a means to examine patch boundaries, or we can look past these symbols for the underlying social status characteristics or lifestyle consumer groupings to define other boundaries for other research and practical issues. To paraphrase Durkheim, we look at social patch clusters not as symptoms but as naturalistic realities of a functioning urban ecosystem.

In a global political economy, the structure for regulating transactions between strangers connects to the flux and flow between biophysical and sociocultural perturbations. However, these patterns often transcend purely local conditions. Olson highlights these patterns and processes. She argues that "there is a tempo and a pattern to urban growth—a dance of larger sweep from generation to generation, as brick rows and marble steps extend over the landscape, as people do their social climbing or take a fall, and move around the corner or up the hill or out to the valley.... City growth is a boom-and-bust sequence of 'long swings,' with eighteen years or so separating the neighboring peaks. This applies equally to North America and Europe over at least two centuries, and to much of the world in the last hundred years."[40] She identifies changes in transportation, political economy, migration patterns as well as the actions of developers, politicians, and other speculators as the drivers of the changing configuration of the city.

Olson provides details on the "swings" of Baltimore's political economy and cultural process: early settlement, the tobacco economy, religious immigrants and disputes, and the need for low-cost labor that brought indentured servants and then slaves from Africa; the emergence of trade and the need for capital expanded new occupations and organizations; the influence of technology on transportation and the means to capture energy to power the emerging industrial sector; fluctuations in war and peace; the changing characteristics of migrants; and the fluctuations of fires and yellow fever are some of the processes that shaped the neighborhoods and overall urban morphology. For our interest in covariation between social and biophysical factors, she provides substantial time series information that identify the constraints and opportunities driving observed spatial and temporal patterns. For example, for the 1802–21 period she notes: "The means of developing and controlling the physical environment increased, yet as I have shown, environmental problems loomed larger than before. The same thing was happening in the social environment: the resources were greater, but so were the disharmonies. The dimensions of poverty seemed to increase. Like smallpox and yellow fever, violence threatened to break into mob rule or anarchy. Everyone saw the visible signs of success and failure all around, and everyone experienced the pressures of boom and collapse."[41]

Later, she notes, floods and drought placed pressure on the social structure. For the 1866–77 period, there were prolonged heat waves. For example, she writes, "During one week in July 1876, deaths reached double the seasonal norm. Heat deaths show the way environmental stress was associated with occupational risk and social assignments. Many of those who died were construction laborers—street pavers, cellar diggers, stone cutters."[42] The great 1904 fire destroyed over fifteen hundred buildings in a 140-acre area in the center of the city. The event generated the demand for better water supplies and treatment as well as a topographical survey that had been proposed for decades.

Patterns of black and white migration in and out of the central city provide the neighborhood identities to be found today. North Baltimore remains mostly white and upper-class, just as it was in the late nineteenth century. In the 1919–34 period, there is the "long and gradual process, by which the suburban counties have shifted from 40 percent black (1810) to 3 percent (1970) while Baltimore City went from 12 to 15 percent to over half, there had been a period of stability and parity—15 to 20 percent black in both city and countryside between the Civil War and World War I. But the trend set in again decisively in the 1920s, as blacks continued to enter the city and whites to exit to the suburbs."[43]

As Olson closes out her analysis in the 1990s, she again emphasizes the environmental influences of long-ignored issues of air and water pollution and the concentration of poverty and poor health in certain neighborhoods. Our task in understanding the ecology of adaptive management strategies is to look for similar patterns in their specific points of action.

We are particularly interested in the work of Bernard T. Bormann and A. Ross Kiester in which they modify adaptive management with their—"acting on uncertainty" approach. This is similar to our HEM strategy, which is to live within the reality of uncertainty rather than expect it to dissolve. As they note, "Forest management outcomes are often highly uncertain, thus a 'best practice' may not exist. We use uncertainty in a general way, to indicate that information on the probabilities of outcomes is lacking."[44]

Their example is the Five Rivers Landscape Management Project, which looks at thirty-two thousand acres in the Siuslaw National Forest in Oregon that contains the Five Rivers watershed management project.[45] It is a landscape that was heavily burned in 1849 and has had modest scattered fires since then. The result by 1993 was a mixture of plantations 5 to 35 years old within a matrix of mature stands 60 to 140 years old. The Northwest Forest Plan in 1993 designated the area as "late-successional reserve," which places the management focus on commercially thinning the rapidly growing and over-dense plantations to speed development of late successional habitat, which also had the effect of helping the struggling industry sector and thereby the local economy. The plan's management/research strategies are "admit uncertainty" and "hedge by diversifying management."

Their plan looks very much like "natural" patches in the eastern forests of North America. Within the plantations, dispersed among the broader older forest matrix, they applied three distinct types of patches— "passive management," "continuous-access management," and "pulsed access management." With these units in place, managers were able to undertake "speed learning" and redefine roles and responsibilities from the learning process. Prediction of management actions is often impossible because of the complexity and nonlinearity of many ecological processes. Bormann and Kiester then go on to show how we can narrow the range of error through a learning process approach. They emphasize the need for the participation of the public in the management of the commons as integral for the learning process the challenge is to find a way to do this effectively.

Here, our adaptation of the HEM to local community pattern and process faces a condition where the resource is physically separate from nearby stakeholder communities, including some very distant but important stakeholders. However, both are very much a part of a resource decision's "virtual" community. The lessons from the three experimental units will need to be vetted in some way for the distant owners in faraway places like Delaware or Maryland with those of the nearby towns directly dependent (like our towns discussed in chapter 3) on certain resource flows. In this case, the chain of supply and demand for certain goods, benefits, and services becomes the thread of knowledge the managers will need to attend to. Here they will need to mix the biophysical constraints with the sociocultural constraints to find the most sustainable option.

EXTENDING THE CAPABILITY OF THE MODEL 237

The manager will need to look up from the soil plots and the history of fire regimes to see how they accord with the divergent burdens and benefits that the close and distant participants will experience. The HEM should be able to identify the nature of the virtual community and predict participant interests and desires as well as radio evangelists can identify the needs and wishes of their distant audience. The challenge for natural resource managers is how most effectively, efficiently, and equitably to exploit the internet's ability to expand and to include community participation by those both near and far from the physical location that so concentrates the managers' attention. In sum, can their field of perception expand to meet the reality of participation available for sharing mutual learning of what the communities know about the "values" of the resource system and what the managers are learning from their experimental plots in the field?[46]

9

In chapter 10 we noted how the Baltimore Vital Signs project was monitoring community health over time. We illustrated how some of their data could increase in utility when organized by the HEM, which permits us to fit together the meaning of the several measures as part of an overall human ecosystem approach. Recently Pickett and Cadenasso have identified how the patch analysis framework can be applied to urban design issues. Grove and colleagues illustrate how patch analysis can advance the scientific capability of urban ecology.[47] We are suggesting that adaptive management experimental strategies, guided by the HEM, may also use the patch analysis framework to gain predictive capability from Durkheim's demographic theory on community solidarity and variations in social density. Today we have technical tools for demographic mapping and long-term indicator measures to predict the most likely future demands on natural resources and institutions.

In the future the questions about environmental issues will demand answers that are larger than those found in the traditional cubicles of intellectual life. We have humbly sought lessons from whatever field of thought that can mix in with other unexpected fields of thought whose mixing might give us a quirkier, but functionally better, model for predicting human ecosystem patterns and processes under stressed conditions. We have looked at a wide range of data, concepts, and ideas. There was selective use of some HEM concepts from chapter 6, such as land, water, and air resources and territory and normative and class and status regulators of the human ecosystem use of space. We have traced how a framework on the structure and function of biosocial patches might connect to social theory of how changes in density of system elements can change the means of adaptation. Durkheim holds that the increasing density of a human population requires a division of labor that creates an organic solidarity based on interdependence in contrast to the mechanical solidarity of smaller communities. We do not subscribe to such a polar opposition. We have seen social patches emerge as something bound together by mechanical solidarity, which buffers many local interactions from the abstractions of system-wide organic solidarity. Social patches become a means to create, recreate, and sustain certain distinct identities of shared difference in lifestyle, language nuance, and religious and political organization.

The dynamics of patches are sustained by a steady flow out of the patch by the early immigrant arrivals and the in-flow of new immigrants replacing the earlier residents. In these flows social institutions like political parties, social clubs, and religious groups adapt and absorb the new population, which sustains the protective comfort functions of the spatial patch. The boundaries of these patches are maintained by some mixture of ethnic pride, food choice, music, lifestyle, occupational status, and stage in family life cycle. Extreme events like fires, floods, and government policies (zoning or "urban renewal") test the endurance and flexibility of the patches, some of which are strengthened and others disappear. In this way the Gwynns Falls patches in Baltimore identified by Cadenasso in 2002 now have a meaning and a metaphor to test questions about the patch origin, structure, and likely endurance.[48]

We have illustrated that many studies of urban ecosystems have built on exchanging metaphors between biologists and social scientists, while others have considered multidiscipline exchanges that cross-species studies encourage, while the patch analysis approach promises a common point of study and opportunity for a full integration of concepts, methods, theories, and metaphors between biology and social science to create a new discipline of "biosocial" science. We do not claim the final answer to any theoretical questions but have tried to indicate that there is a fertile intellectual basis to link biophysical and sociocultural elements. In this sense, the patch approach may be able to ground the utility of the HEM, while adaptive management strategies joined to demographic tools of analysis increase the forecasting capability. The next chapter will lead us into some of the certainties and uncertainties facing human ecosystems in the coming decades.

T·W·E·L·V·E

Leaning Forward

Future Challenges to Human Ecosystems

1

There are many ways to examine human ecosystems. The most enduring results occur when social and biophysical professionals share a common language for describing and analyzing the system and when there are core concepts whose meaning and units of measurement are mutually respected. Our discussion of the origins, development, and applications of the HEM suggests that such hopes are both organic and emergent and always in dynamic tension. The HEM adds depth to our ecological thought and practice because there is a realistic script whose concepts link the regularities in human behavior with the regularities in biophysical behavior.

Our model organizes the lessons learned in the field and ways that those lessons can be applied to the management of future ecosystems. The model describes human ecosystems and identifies the critical variables and connections that bind or fragment the integrity of these systems. The model is a base for linking the many disciplines, theories, methods, and measures seeking to understand complex environmental issues.

The world today is substantially different in many ways from the 1970s, when we first tried to develop a framework for combining biophysical and socioeconomic factors for understanding human ecosystems. We can point to many specific ways in which environmental quality has improved—some species of wildlife have been restored, awareness and some action on climate change has been made, and some watersheds have been revitalized. On the grand scale, however, the trends for many environmental and social issues are less hopeful—as we described at the beginning of this book. Air and water are still being polluted, virgin forests are being converted to other uses, large numbers of the poor still remain poor, and the rich still get richer, often at the poor's expense. There are several trends whose future consequences for human ecosystems are fairly predictable—growth in human populations; the urbanization of our species; aging populations in wealthy countries; large-scale migration of peoples from poorer to richer economies; climate change; deforestation; energy pollution; loss of cropland; flooding, fire, and other extreme events; diminished supplies of potable water; and more. The HEM identifies the most likely possible connections between the relevant trends and their interdependent connections and permits a general estimate of the lag between these events and the most likely biosocial responses.

Trends in *social institutions* may be overlooked yet critical to the issues that concern many resource managers and conservationists. We think that three are more important than are usually recognized and will be key factors in the future. These are: *justice* institutions, particularly the mechanisms of property rights and responsibilities where the norms of hierarchy sustain growing disparity in wealth between and within countries; *sustenance* institutions, where capital exerts hegemony over labor and consumerism drives growth or decline in the political economy; and *faith* institutions, where cultural resources of organization and belief give legitimacy to participatory ecosystem policies and practices.

2

We begin with an examination of property rights institutions and the norms that regulate access to and responsibility for natural resources. In particular, our interest is in the human ecosystem consequences where large amounts of wealth are held by a relatively few persons. A recent

Oxfam study detailed the disparity. Patricia Cohen summarizes some of the findings: "the richest 1 percent are likely to control more than half of the globe's total wealth . . . the 80 wealthiest people in the world altogether own $1.9 trillion . . . nearly the same amount shared by the 3.5 billion people who occupy the bottom half of the world's income scale." To focus our analysis, we use three historic examples—an effort at botanical piracy, creation of utopian community, and a massive forestry experiment. All three examples have to do with the Amazon basin. The outcomes found in each of the projects depend on a great disparity in the distribution of wealth that often encourages a level of hubris, whereby the person treats his or her wealth as a sign of virtue that entitles him or her to do "good" things "for" the less fortunate.[1] We will use these examples to gain a better understanding as to how good intentions fall victim to forces that turn them into a series of unintended consequences.

The HEM framework has *wealth* joined to four other hierarchies—*power, prestige* (status), *knowledge,* and *territory*—that regulate access and motives to "own" certain desired resources of human ecosystems. We might assume that the very wealthy, as they become more wealthy, might tire of endless consumption and reach out to have more depth of meaning in their lives. One action might be to gain power to control the government and through that connection seek to manage the biophysical resources and in consort manipulate some socioeconomic and cultural resources. In this way the wealthy donor consolidates power and a pathway to joining the higher prestige and status represented by "old wealth" families. The history of American crime families suggests that as these families accumulate wealth they move to manage a Tammany Hall or similar political machine to demonstrate and to sustain their wealth. This base is used to fund their children and grandchildren with Ivy League degrees and prestigious positions as chairs of art and culture organizations and museums. This is one route of "mobility" for the well-to-do. However, another is to create one's own kingdom.[2] It is this type of hubris that we will consider as a future archetype that will challenge conservationists and natural resource professionals.

The first example considers the great economic, environmental, and social costs from the biopiracy work of Sir Henry Wickham in the nineteenth century. Wickham stole seventy thousand rubber tree plants from the Amazon for the "prestigious scientists" at Kew Gardens, who did not know how to grow them in Britain's Malay colony and were unwilling to listen to a commoner like Wickham.[3] Ultimately the rubber trees were adapted to the new habitat, and his biopiracy expanded the power of Britain and brought down the price of rubber for customers in the wealthy countries, but the poor villagers and workers in the Amazon watershed paid the high costs for these benefits to others. These patterns of globalization are not unlike those of the twentieth and twenty-first centuries in terms of their human ecosystem consequences. Present and future interest in medicinal plants, parts from endangered wildlife, exotic pets along with timber, energy, and minerals are likely sources of property stolen from indigenous communities and converted to property of a dominant colonial power. The theft of the rubber tree seeds was enabled and approved by scientists working for one of the leading botanical and biological research organizations in the world. There is no evidence of any remorse about the theft of the seeds, nor any suggestion that the botanists cared about or considered all of the biophysical and sociocultural consequences of their actions. Their emphasis was on serving the powerful large-scale colonial enterprises. Similar challenges will be faced by twenty-first century science. The HEM might guide scientists to consider the human ecosystem consequences of such harvesting and to find ways to alter or to mitigate or to stop these activities. Other than the economic gain, there are issues of property rights, equity in distribution of burdens and benefits of the resource actions, damage to the structure and function of the ecosystem, and the long-term consequences of maintaining outsider hegemony over local people.

Our second example takes place in the early twentieth century and in the same region of the Amazon where the rubber tree biopiracy took place. It involves the vision of the first true industrial ecologist, Henry Ford, and his utopian industrial town Fordlandia.[4] Ford thought that he could "manage" human and biological systems in the ways he had done in industrial factories. His efforts at "managing" human and biological communities, whose essence he did not understand, created huge economic, social, and biological failures. Ford and his professional consultants had little knowledge of the local people but assumed that they would enjoy a replica of a Midwestern U.S. town with all of its constraints on working times, drinking opportunities, and revenge between clan groups. On the biophysical side, the unique wet/dry cycle driving the hydrology of the Amazon's ecology and its influence on the characteristics of plant species and soil conditions interacted with the limitations of introduced species for this environment and their lack of resistance to indigenous insects. None of these possibilities were considered. Ford's consulting professionals failed to consider, let alone prepare for, all the complexity required for success from such radical biological and social interventions.

Our third example is set in 1962, when a shipping and real estate billionaire, Daniel Ludwig created an even larger project in the Jari River area of the Amazon. It emphasizes that dreams about society and nature in exotic locales are like Indonesian shadow dancers, where the image plays on the misperceptions held by the wealthy outsiders who plan to reorganize human ecosystems to make them more efficient or effective. Driven by a faith that the world was going to face a timber famine, Ludwig gained rights to eight thousand square miles on both sides of the river. His workers stripped 250,000 acres of virgin native forest, to be replaced with a fast-growing species, *Gmelina arborea* from Nigeria. When this species did not work because of inappropriate soil conditions, other exotic tree species were tried. These failed as well. As Jackson notes, "after fourteen years, Daniel Ludwig gave up, losing an estimated two thirds of a billion dollars to the same mistakes Ford had made."[5] A Brazilian group bought him out and reported likely profits in the future. Jackson reports but does not measure the untold damage to biophysical elements of the ecosystem for this huge area in the Amazon watershed and leaves unmentioned the damage done to the indigenous peoples who lived in this region.

Certainly there will be many large landscape tracts for the wealthy to own for conducting social and botanical experiments. Though the wealth of the new rich will come from digital enterprises rather than the mechanical enterprises of the industrial barons, they will continue to follow the old practices of "doing well by doing good." One outstanding example of the "new" rich is Larry Ellison, CEO of Oracle, who in 2013 was paid $78.4 million ($37,692.31 per hour for a forty-hour week), though the profits of his company were fairly flat. He managed to persuade the government of San Francisco to host the finals of the 2013 America's Cup sailboat race, for which the attendance was low. In February 2014, the San Francisco Budget and Legislative Analyst reported "the daunting reality that hosting the event has left San Francisco nearly $11.5 million in the red."[6] In 2012 he bought the Hawaiian Island of Lanai. The island is 141 square miles, 18 miles long, and 13 miles wide with a population of 3,200. Ellison has spent several hundred million dollars to acquire nearly 97 percent of the island's ninety thousand acres.[7] Its primary economy was the production of pineapples and sugar; now it is tourism, and its future is uncertain.

The challenge for natural resource professionals is to make the complexity of human ecosystems an empirical reality for their wealthy clients. Property rights institutions cannot be disengaged from responsibilities to the land, the people, and their future. Natural resource professionals should be reasoned critics rather than enablers who give their nodding assent to all the ideas the rich and powerful may come up with. The HEM in its present form can be a useful tool for such critiques. Without these critiques it is certain that the new rich will let their hubris overrun their common sense and seriously damage people and landscapes, as was done in the Amazon. Perhaps the challenge is to develop a preemptive restoration ecology that plans for the many anticipated large-scale farms in the Amazon and other frontier wildlands.

Leaving the ultra-rich aside, there are the moderately rich who are laying out enormous soy farms in the western Amazon, which Stephen Ponder and his students are studying for their sustainability. These farms are likely to be an expanding future implant in the wild jungle. As Ponder notes: "By 2050, a global population of nearly 10 billion people will require roughly 70 percent more food. Either we need to get more food out of the land we already farm or we need to farm more land."[8] His research is measuring the ability for these mega-farms in the Amazon to produce at a sustainable level of output. He notes that natural science alone cannot answer the many questions about the impacts of these large soy bean farms and finds that he is more and more talking to economists and sociologists. In short, he is looking for a HEM that helps farmers and agronomists adapt to these new realities.

Clearly, attention to the varieties and applications of property institutions and social hierarchy will be a necessary component for understanding these future conservation issues. The HEM provides a context for anticipating the likely connections between ownership rights and responsibilities and their underlying mechanisms of change and stability.

3

We now look at sustenance institutions in general and the socioeconomic resources and ecological realities to be anticipated in the transitions from a producer to consumerist political economy. Decision structures for policy and management of natural resources and control of environmental impacts now have an amazing supply of information without the filter of knowledge. Social media technologies now give legitimacy to previously unheard opinion. The supply of so much information may find its positive gains overwhelmed by the noise of so much information. John Berger observes:

> We are living today in a culture of information. I use the word 'culture' in its anthropological sense; the information-culture has in practice no place for cultural heritages of any kind. It stimulates calculation but consistently discourages reflection. Thus it substitutes information (and misinformation) for knowledge or wisdom. This is alarming, yet it's a culture that sooner or later will spin out of control; it will not endure.
>
> I start here so we may remember that knowledge, as distinct from information, always allows for and reckons with the unknown. As bodies and disciplines of knowledge extend, so does the extent of the unknown. Perhaps the relative dimensions of the two is a cultural constant. Yet the frontiers between the two (the known and unknown) are being continually contested.... Many forays are scientific, and other more intuitive but no less important ones are made by mystics and artists. When an insight brought back from

an intuitive foray seems to stand up, hold water, or prove its paces, we are in the face of what can be properly termed an original work.[9]

Berger is telling us that simply accumulating more information (and misinformation) does not give us knowledge about the likely future consequences of present actions. Indeed, what is needed is to sort the flood of information into functional clusters that can creatively lead to knowledge. The ecological mantra that everything is connected to everything else may be so, but it is not very useful in actually solving problems. We need a guide that points us to the most critical connections that minimize choice and maximize the potential for solutions for renewing ecosystems. We may be certain that Ford and Ludwig had large amounts of information, but much of it was the wrong sort, and there was no knowledge as to how their several actions would cumulatively interact and thwart their desired outcomes.

The 2008–09 global financial recession and the increased global unemployment that continues have been swamped with piles of information and very elaborate metrics seeking to forecast cause and consequence of various policy actions. For the wealthy countries, the one certainty is that the "health" of the economies requires considerable levels of consumption. It does not seem to matter what people consume, just that they go out and buy things. This, of course, means that what is bought must be quickly discarded so that replacements can then be purchased and keep the workers in Bangladesh sweat shops sewing, keep the ships bringing the goods from port to port, and keep the low-priced items gliding through the Walmart checkout. Of course, this also means the expansion of landfills, recycling services, and the supplies of materials to produce the goods.[10]

A tremendous amount of research is generated to target niche markets effectively for various products. This is financial investment that does not go to understanding microbes and viruses and their possible management. Rather, it is research taken from credit card purchases, smartphone calls, and texts that supply information about how to sell something to a targeted population.[11] New wealth is not created by combining the usual triad of capital, land, and labor. Rather, it is information not to increase productive efficiencies but to increase consumption expansion.

Clearly, all the environmental rules and regulations, and monitoring and conservation practices cannot withstand such a system that is essentially based on waste. Studies by psychologists find that increasing levels of consumption do not generate increased levels of satisfaction or even joy. Of course, we need paid employment to access the necessities of living. Yet to have the primary driver of this employment be based on endless acquisition of things and the rapid cycling of these purchases to landfills seems an unsustainable and unsatisfying quality of life.[12] This lack of knowledge about the reality and possible alternatives to consumerism seldom appears in the gloomy environmental reports that rightfully get at immediate issues but always seem to ignore the underlying causes that ultimately will undermine present good works.

Indeed, most analyses of the major environmental ills from loss of farmland to global climate change seldom consider that the underlying driver of these problems is an economy whose "health" is dependent on unending consumption rather than the joining of land, labor, and capital to produce new actual wealth and to make the daily life for ordinary citizens more efficient, effective, equitable, and sustainable. The historian Jerome Steffen argues that the American dream is based on three myths: "personal and national prosperity is a direct result of the freedom that Americans have to pursue their own destinies"; "material resources in America are infinitely available to those who work to acquire them if the system is allowed to function properly"; and "pervasive influence of manifest destiny and mission . . . has assumed that its way of life . . . is the only one that has any real claim to legitimacy." The abundance of land resources fueled the belief of the three major organizing myths. Steffen cites the analysis of Walter Prescott Webb in noting the four-hundred-year boom from 1500 to 1900 during which "the person to land ratio of 26.7 per square mile dropped to 4.8 people per square mile at the start of the twentieth century."[13] Steffen argues that the faith in abundance is cyclical in its restoration of the myths where hopes of a new future fail to be sustained and the next cycle of technological innovation or scientific management will rebuild the resource base of the myths. He suspects that the digital boom will have a similar boom-bust phase where reality will confound the achievement of easily gained and sustained material abundance on which the three myths depend.

The work of Steffen and other historians and geographers can help us to identify which stage in the cycle of our operating myths we find ourselves and which variables are most crucial in determining the timing of the flux in a belief system and under what conditions they shape attitudes and behaviors regarding the finiteness of material resources.[14] This reality of perception about the necessary and sufficient factors driving a political economy will challenge the next stage in the evolution of the HEM. The model can help to filter and organize the mountains of information and force a closer connection to knowledge and thus a more realistic appraisal of available abundance. Still, the behavior of restoring, reusing, and refusing material goods cannot be accomplished without attention to how large numbers of present and future populations will have the employment or other economic means to support the material consumption needed for a "healthy" economy. Though human ingenuity may find a way around the constraint of a diminishing supply of many material resources, the quality of life will be significantly diminished.

4

Our third focus of attention is how faith institutions and cultural resources will shape the likely futures of human ecosystems. The institutions of property identify, regulate, and order rights and responsibilities of the owner for specific resources, whereas sustenance institutions regulate social behavior in terms of the nature and types of available material resources and the characteristics of access to these material resources. Both sets of these institutions have a strong sense of rationality underlying their organization. They seem to make routine those regular reoccurring events faced by persons and their human ecosystems that appear orderly and rational. Who owns these resources? Who gets to use them? Who maintains them? To answer these questions there are mechanisms like laws and courts, markets, rules of trade, and perhaps most importantly norms for positions within hierarchies of differential access to prestige (status), wealth, knowledge, territory, and power.

Faith institutions organize issues that have an internal rather than an external and empirical rationale. Faith institutions provide answers for the un-answerable about life and death, good and evil. One believes because one believes, and holding that belief makes it real. Faith institutions are based on emotional proofs and, therefore, strongly shape human response to nonhuman nature. There are many essays and books and speeches on the powerful connection between a given body of faith and its shaping of the behavior of believers toward use and constraint and awe in regard to nature. Still, much of the "proof" has been more narrative than empirical. One of the major challenges to future HEM applications will be unpacking the relative importance of differing faiths regarding the structure and responsibilities of human response to certain ecological realities. We know that some Christian Evangelical faiths deny the reality of evolutionary theory and the reality of climate change. We read the encyclical letter from the first Jesuit Pope that accepts both and urges action. The paradox of globalization and ethnic fragmentation running parallel to one another seems a challenge that will play out in future practices of human ecosystem professionals.

A classic social science approach to understanding the relation between institutions of faith and socioeconomic behavior is the work of Weber on the relationship between the Protestant ethic and capitalism.[15] Like other scholars in the late nineteenth century, he was trying to understand why industrial capitalism emerged in Western European countries rather than great nations like China, Babylon, and Egypt. These nations had many of the technologies and capitalistic adventurers in trade and warfare who sought wealth. But none of these nations developed the rational, quantitative systems of deferred gratification observed in the rise of modern capitalism. Weber argued that the attitudes and visions of a culture were likely to be the primary drivers of socioeconomic behavior. For the purposes of the social ecologist, Weber's work helps us to understand the importance of faith in shaping future behavior and how variations in practices, belief, and organization of a given faith shape behavior in the profane world.

Weber's method is akin to that of a wildlife ecologist who collects scats from their focus animals to analyze their diets and to predict likely future behavior. Weber could not go back in time or travel to faraway places, so he examined the leftovers of human life, the intellectual "scats" from document remains, sermons, texts, reports of observers, reports by historians, statements of goals and rules, and visions of the faith and of a worldview. With this information he compared "ideal" types to stand for a given set of faith and behavior to define the "ideal" types of Calvinism that serve as a sum of the beliefs and behaviors of a collection of believers.[16] Weber's human ecology is an analysis of the complexity of connections and interactions that did not all link up in other cultures but with the authority of faith formed the bond that led to an unintended consequence—the emergence of capitalism in the Occident.

Weber's writings give an idea as to how we might interpret the importance of faith in shaping a response to the equally dramatic shift from an industrial production system to a global consumer economy that is driving socioeconomic change in postindustrial societies. We want to understand the institutional forces that will be shaping human relations toward one another and our shared earth in the coming decades.

5

The importance of faith institutions for conservation may be experienced when we explore parallel empirical realities where different faiths produce different human-nature interactions. We will not concentrate on the negative practices that some institutions of faith can engender, nor will we look at the discussions as to why this or that faith is more "eco-friendly." Rather, we will compare two nations—Bhutan and Nepal—that were closed off to most of the world until the 1950s and that are also squeezed between two of the world's fastest-growing economies and human populations—China and India. Bhutan is a predominately Buddhist country, and Nepal is predominately Hindu. Nepal is larger in territory (56,827 square miles) compared to Bhutan (14,824 square miles) and much larger in terms of population (29,391,883, leading to a density of 531 per square mile) than Bhutan, which has 706,427 persons for a density of 48.8 per square mile. Both countries reflect adaptation to the demands of living in the highest mountains of the world with melting glaciers impacting a predominately rural agrarian economy. The demographic trends in both countries are toward a more urbanized, non-agrarian economy. In terms of religious faith, Bhutan is 75 percent Buddhist and 25 percent Hindu, and Nepal is 81 percent Hindu and 11 percent Buddhist. Neither country was a colony of a European power. Although their actions were circumscribed by treaties with the British, they were never under occupation, as was the case with India and Pakistan.

These two countries provide a good comparative base with their similar challenges of deforestation, rural poverty, and the press of international tourism. They both face similar challenges to their forest ecosystem practices and from the intervention of international advocates whose resumes of strategies, failures, and successes may predict likely outcomes for future environmental issues.[17]

We compare the two in terms of different approaches to meeting ecosystem challenges: Bhutan in terms of mixing faith with conservation and Nepal in terms of using organizational strategies, particularly community forestry and its national organization, the Federation of Community Forestry Users Nepal (FECOFUN).[18] Both countries have valuable data sets collected by their respective government agencies, international agencies, and various private donor groups. Future researchers have the opportunity to use the available data and to measure the consequences of varying interventions and the relative impact of justice institutions, sustenance institutions, and faith institutions on specific critical resources. Some background is necessary.

Worldwide concern about the deforestation of tropical forests and the loss of biodiversity arose in the 1970s and 1980s. Since traditional responses of government centralization and regulation of forests did not work, conservationists were forced to link their ideas of nature conservation to the desires of local communities for "development" and participation in the planned activities desired by conservation professionals. However, the reality on the ground demonstrated the limits of such cross-cultural exchanges. Conservationists want villagers to protect the tiger and the rhino, and the incentive is that they will obtain the material goods, like those in wealthy societies. In the early stages of "conservation development" projects, there was a tendency to expect rural communities to do what the foreign conservationists wanted rather than what the local groups might desire. Nor did such efforts consider that over time communities would have evolving goals, such as getting out of subsistence agriculture.[19] In many cases the projects did not succeed in the various goals of saving the forests, the tiger, and the rhino or of improving the livelihoods of local people.

The stories of success and failure in projects to conserve forests and biodiversity have substantial sources of data and findings for learning what works, what does not, and why, with various policy and practical interventions serving as the social experiment. Yet there is little cumulative analysis of the driving forces leading to specific outcomes. This is because projects have short half-lives of three to six years in duration, and both the external and internal professionals carrying out the projects need to leave for their next projects. Nor are there any incentives to cross project boundaries for cumulative learning. So the Japanese projects do not share their monitoring data with the British or Australian or American projects. Consequently, there is limited means for cross-referencing the findings with other similar projects. There is no institutionalization of the findings that cross agencies, NGOs, and so on. Meanwhile, other environmental challenges such as climate change, loss of cropland, and air and water pollution will crowd into the future. Project managers could benefit from the use of the HEM to explore cumulative findings and treat them as present and past "experiments" on "solving" environmental challenges (in the case of Bhutan and Nepal, forestry and biodiversity questions). The environmental professional can use the model and these findings and data to ask about what specific drivers shaped the consequent outcome of the prior intervention. For instance, what particular associations were most essential in shaping the observed outcome? Were they variations of faith institutions? natural resource technologies? cultural resources? property rights institutions? biophysical resources? specific policies of government institutions? Our comparison of the different pathways in Bhutan and Nepal gives a hint of what might be done with particular attention to the role of faith institutions, which often get overlooked in the desire to find a quick technological fix that fits within the time limits and other constraints imposed by the project design.

Both countries have only recently been exposed to the global economy. They still exist within traditional rural social patterns but have entered the new world in different ways. Bhutan has been much more cautious—government employees must wear traditional attire, television was permitted comparatively late in the twentieth century—while Nepal became an early adopter of foreign styles, products, and temptations. However, in both countries, outside of the cities the rural ways predominate. Both countries share a common geography and resource base but have very different meanings guiding the use of their ecosystems and the sacred role they play in the daily lives of the people.

The other side of the conservation/development strategy is what the highly trained expert brings to the environmental challenge facing the high Himalayan communities of Nepal and Bhutan. In general, the record of most introduced changes, whether technical or cultural, reflects limited long-term success.[20] The earlier cases about rubber trees, exotic pulp wood, and planned natural resource-based communities in the Amazon are confirming examples of the limits and liabilities of expert interventions. There is little thought about the unintended consequences of imposing strongly held perceptions on biotic and social ecologies in unknown "foreign" places, which often incur massive social and economic costs.

We have cases from the nineteenth century, the 1930s, the 1950s, and the 1980s, where similar errors were recycled by rural development projects. Our hope is that the conceptual lens of the HEM can minimize the frequency of undesired consequences. Its future use can be both preemptive and proscriptive in helping local people sort out their respective adaptive strategies. Hence we direct our attention to faith and the cases of Nepal and Bhutan in creating their own solutions from within their own national perceptions. In these high-elevation nations, we see a distant reflection of the responses found in North America and elsewhere.[21] The outside professionals serve as facilitators rather than as directors in the search for a sustainable quality of life in a challenging physical environment. The challenging physical environment of the high Himalaya simply sharpens this reality.

6

The power of the high Himalayan natural environment on the lives of people of Bhutan and Nepal is best experienced in the dramatic cycles of dryness and wetness that regulate much of the daily life and larger patterns of survival in the region. During monsoon season, all activity, from talking to crossing a street, is determined by the volume and velocity of rain. The sky is a waterfall that envelopes one in rain so thick, so persistent, and so determined, it seems as if the Indian continental plate has been raised up somewhere near Sri Lanka and the whole contents from the Bay of Bengal is being dumped on everything in sight. This is not the little constant drip, drip, drip of London or Portland, Oregon. This is rain of such singular intent that one can have no doubt that it stems from gods of great strength. It is also the time when there are fewer development advisors to be seen giving advice to villagers.

Grand and wonderful as other mountain ranges may be, they do not have the startling rise of the southern Himalayas from near sea level to the highest place on earth. The North American Rockies, for example, rise from a mile-high plain. Nor are the Rockies as thickly populated with human communities, many of whom live in a temperate climate although the geographic context is the tropics. For example, some of the highest-altitude permanent communities are the nearly five thousand Dolpo-pa who live in the valleys of western Nepal at fourteen thousand feet and graze their yaks at sixteen thousand feet. Here a slight shift in temperature or rainfall can have dire consequences for the resident peoples.

The Himalayas stand alone, marked with geological and human history at such a grand scale that we lose our voice. Words rattle around in cliché. It is not talk but feeling that gives one access to their meaning. Anyone standing at the nearly flat base in the Terai can "feel" them rise up and up in such a determined pitch that one feels the need to step back just in case they decide they have reached a tipping point and are about to topple down. Mountains as grand as the Himalayas do not exist in terms of some technical definition, nor as an example of some large biogeochemical experiment. The many highland human communities that span from North to South and from East to West fill in a series of diverse ecologies where people are as much a part of the persistence of the landscape as the rocks and plants and geological forces. In a very realistic way these are not landscapes carved from the "wilderness" ideal of Western countries. Indeed, they are thickly peopled landscapes where people coexist with monkeys, tigers, and takins. They provide lessons of change and adaptation for those living in less demanding environments.

The natural environment demands, powers, and sustains the reality of the religious faith. Its very drama is a fit setting for the miracles and rituals that bind the believer to the reality promised by the faith. Uncertainty about the rainfall, the seeds, landslides, manure, and markets limit a desire for risk-taking by subsistence farmers. Their best hedge is on looking to the past and following traditional practices that are firmly based on generations of empirical experience. For all the unknown and known challenges remaining for the farmer, the practices and rituals of faith provide a sense of control over the remaining uncertainty. The subsistence farmer is realistic rather than stubborn. The farmer requires a conservative caution toward adopting many of the suggestions made by foreign experts that have neither locally empirical knowledge nor sacred sanctions to follow.[22]

Poor countries and regions often have landscapes of great wild beauty and seemingly empty spaces that contrast greatly with the highly rationalized landscapes of wealthy countries. Not too surprisingly, these great natural landscapes probably still exist because the human residents remain at subsistence or greatly reduced levels of income. The necessities of poverty and harsh growing conditions drive very different notions of what a tree "means." To the subsistence farmer, it is not beauty or board feet but rather a means for reciprocal connections in community and family life. Often such trained visions encourage the farmer and the development professional to talk past one another.

Consider that a forester from Austria or Maine comes to Bhutan and climbs up into the hills and is excited by all the old growth hemlock that looks just like the hemlock back home. This forester feels comfort in finding something that is so familiar and for which there are programmed "solutions" for this tree to "help" the Nepali or Bhutanese farmer. That Austrian or Maine forester looks at the tree with a capacity that is so well trained that it becomes a liability in judging the meaning of a specific tree within a different cultural context. In Nepal or Bhutan, a tree may be seen as part of a sacred community or a legacy from an honored relative, or it may be capital available for the bride price of one's daughter. However, in Maine or Austria there is the strong sentiment that God gave the wonders of the earth for humanity to exploit for its own interest. It is a commodity that generates income or is put away in a sanctuary to serve tourists. It is seldom seen as part of our human community, as our sisters and brothers, worthy of mutual respect. The tree to the villager shares both mystical and material (sustenance) dimensions. It is embedded within a faith, and therefore its use is not primarily a market-driven one but rather a gift or a request for mutual help toward a fellow member of the human ecosystem that includes plants and animals as well as people.[23] In this situation the institution of faith is at the core of factors affecting human-nature transactions. The natural resource professional needs to learn how to work within this frame of reference, rather than against it.

7

We have suggested that there are empirical bases for understanding the future by treating some past policy or management practices as *ad hoc* experiments. Like the subsistence farmer we are leaning forward to have more effective, efficient, and equitable outcomes from our decisions about fixing future environmental challenges. We have considered two similar and yet very dissimilar countries as a comparison for thinking about the role of institutions of faith in regulating the ecological dance. We have discussed the impact of the Himalayan biophysical reality on human ecosystems and how such extreme conditions give useful insight for systems in less dramatic settings. This final section will give a short snapshot of some basic elements of faith by Hinduism and Buddhism and their consequences for environmental sustainability.[24]

Our discussion begins with Hinduism, which emerged as a religion 4,500 years ago in the Indus River valley. It is the oldest of all world religions.[25] Its present number of followers is 800 million, most of whom live in India and Pakistan with significant populations in Sri Lanka,

Indonesia, and South Africa. About 1.4 million followers live in the United States. It has multiple gods and goddesses but no single deity like Islam or Christianity. Nepali people, with a bit of humor, sometimes say that there are "enough" gods for everyone to have his or her own spiritual advisor. Certainly there are enough rituals and special holidays to sustain the reality of the faith for individuals and communities. Its symbols, temples, and scriptural passages reflect an earthy agrarian world perspective.

Jagannath Adhikari in a forthcoming book reflects on his family's agricultural life in Nepal: "Farmers were constantly worried about what their resources would tell if they were not used properly. While doing farming or any other work related to nature, [a] trilogy of concept(s) 'dharma' (merit), 'paap' (sin) and 'karma' (deed) were invoked, which were based on traditional worldview not only of Hindu religion, but also on various local faiths like animism and Buddhism."[26] He argues that this worldview guided the farmers to perform in an ecologically sound way and to support ideas of social justice in the communities where they lived.

Adhikari and others argue that the precepts of Hinduism have been overwhelmed by ideas of the green revolution that do not fit small farmers in difficult growing situations. Modern ideas about farming have pushed [them] aside with the loss of flexibility, equity, and sustainability.

Adhikari reports that traditional agriculture was effective in preserving biodiversity: "there were as much as 2,000 landraces (varieties) of rice alone in Nepal growing in areas from 60-meter altitude to 3,050-meter altitude. Furthermore, another study revealed that of the known 1,800 varieties of indigenous landraces of rice including wild species in Nepal, only one variety is cultivated in the Tarai, where rice is widely grown.... Modernization process which led to erosion in the tradition of rituals and faiths has also led to this decline in indigenous landraces and local species of plants."[27]

Adhikari and many other observers note that traditional agrarian practices sustained by the Hindu faith are at risk of extinction. The wave of Western technology may reach its peak in the coming decades, forcing the children of local people to leave the hills for the cities. This central tendency has not been as forceful in Bhutan. Further in Nepal, there are countervailing forces in the expanding numbers of community-based natural resource systems. Here the Nepali farmer is using some of the organizational ideas of the foreign advisors to sustain tradition and push back against the erosion of traditional ways. We noted earlier that centralized forest management could not control the decline of forest health. The nationalized forest land was once the common land of the villages, and when the village ownership was taken over by the central government local people had few incentives to look after the forest resource. "Access to forests is important because they are central to farming systems, and more than 80 percent of Nepal's population of 23 million is dependent on agriculture. Fuelwood supplies over 75 percent of the country's energy requirements. Moreover, forests play a critical role in the agroecological cycle of cropping patterns, animal husbandry, and forest products that sustain agricultural production."[28]

The new community-based forest management system continues to expand and seems likely to be the predominant mode of forestry practices in the future of Nepal. Here property institutions are helping to sustain the values of faith that encourage sustainable use. There is a reciprocal reinforcement pattern where the values of faith and the structure of property institutions may protect traditional values and practices. In 1995, the Federation of Community Forestry Users Nepal (FECOFUN) was organized to strengthen the rights of local communities to participate in policies affecting their forests. In 2013, there were some 8 million forest users involved in the organization and fifteen thousand community forest user groups and other community-based management groups such as leasehold forestry groups and religious forestry groups. Approximately thirteen thousand of these groups are affiliated with FECOFUN. Over four million acres (39% of Nepal's forested lands) are managed by these groups, and 40 percent of the Nepali population are members of community forestry user groups.

On its website, FECOFUN describes its activities: "A national advocate for forest user's rights . . . effectively lobbied government officials and politicians, organized marches, demonstrations, and signature campaigns and filed court cases. It has been successful to unite the forest users across Nepal and acting collectively to gain and retain users' resource management and use rights . . . made possible to engage in debates about community forestry's futures in local, national, and international arenas . . . has been (involved) in technical support for user groups in the areas of forest resource management and empowerment." Its goal is "to develop means of livelihoods and to generate opportunities to reduce poverty through forest management and utilization of resources, abiding by the approach of consensus making. And to emphasize the participation of all sections of community and ethnic group decisions."[29]

Nepal's mingling of cultural resources, such as belief and organization with institutions such as faith and property, may be a pattern for encouraging widespread public commitment to solving environmental issues. Here the strategy is to have a strong external force for protecting and sustaining latent internal stewardship values to direct individual and community behavior. Future challenges for environmental professionals will be the mobilization of such public action toward balanced use of nature, and lessons from Nepal may prove critical to others as well. The conceptual means for asking such questions is given by the HEM.

Buddhist Bhutan offers a different but complementary strategy for mobilizing widespread public participation in ecosystem stewardship. Buddhism was greatly influenced by Hinduism in that there is no god of judgment, the daily actions of people have spiritual consequences and reincarnation, all dictated by the wheel of life on which enlightenment ends the cycle of death and rebirth and liberates the person from suffering. As with Hinduism, Buddhism's 350 million believers live primarily in Asia. Buddhism emerged 2,500 years ago. Its teachings are based on the miraculous birth of a high-caste boy—Siddhārtha Gautama—in Lumbini, Nepal. As a young man of twenty-nine, he traveled throughout the subcontinent, lived in poverty among strangers, and meditated on the suffering of humanity. The Buddhist scriptures report many miracles associated with the presence of the Buddha. Most of the miracles reflect the human condition as influenced by the environment where the dramatic rise of the Himalaya Mountains from the plains forces humility and miracles to arise.

It is important to know that even though there is a core of thought and philosophy in the spiritual faith, there are two historically different variations in the faith: the "old wisdom," or Hinayana, and the "new wisdom," or Mahayana. The Theravadins, a subgroup of Hinayana, dominate in East India, Thailand, Sri Lanka, and Myanmar. A Tantra spillover from Hinduism influenced the art, dancing, and poetry of the Mahayana branch of Buddhism in Tibet, Bhutan, China, Japan, Java, and Sumatra in Indonesia. One gets more active rituals such as dance and music expressing the ideas of the Mahayana base in Bhutan. It is an active faith. The dances and music are part of the school curriculum and sustain the universal connections between people and nature in Bhutan. We mention these geographic divides to indicate that though the values of organized faiths structure human behavior, their guidance is not absolute but often reflect external biosocial environmental features. Still, the archers in Changlimithang Stadium in Thimphu and the black hat dancers in Kanglung are expressing different dimensions than Hindu rituals and celebrations in Nepal.

The diversity spun off from the two main branches of Buddhist thought reflect a determination that its core values adapt to the realities and unique dialects of a given place while remaining true to a central body of faith. There are variations by country where Japan and China have very different practices from one another and from those in Nepal and Thailand. In rural areas, there are variations by elevation. As one moves from the plains of India to the higher hills, the influence of animism and other traditional beliefs and rituals increases as the altitude rises. In short, the faith is responsive to the biophysical variations as well as the sociocultural variations it encounters. Hence, unlike bodies of faith dominated by a central authority with a substantial bureaucracy to sustain it, the "localism" of Buddhism gives the social ecologist an understanding of the underlying values that guide the local meanings of human and nature relations. It is a true key to the values that guide human action in such matters. So our forester from Austria or Maine would be wise to begin with the faith and the rituals, dance, poetry, and music of the people before considering the trees and what their market value may mean.

In Bhutan, the *dzongkhags* (provincial centers of sacred and secular authority) combine religious, political, and economic dimensions of life for local peoples. These social structures reflect the central unity of Bhutan—faith, nature, and government (particularly the royal family) are the overlapping circles, the binders of social life. Buddhism in Bhutan is very much influenced by its social and physical environment. It is awash with scholars, interpreters, and religious visionaries, each of whom goes mostly in his or her own direction. Like most faith explorations, these have gone on for a long time. In the present case it is an exploration that has been done for over two thousand years, always sifting and seeking and testing from older texts to create a new ecology of explanation that comforts the present with what seems new but feels familiar. Yet each explanation overlays other sets of meaning that also explain the structure and process of a given place that seems sacred even though there is no means to empirically test such philosophical assertions. In short, like money, it is a human fiction invented to help cover our confusion.

Hence the ideas have the power to direct and to sustain certain patterns of human behavior. Buddha was concerned with the pain and sorrow of existence and how to overcome them. He did not seek an answer in some outside force, some supreme god. Rather he found "solutions" to be in the pathways of human thought. If we control the desire driven by "self," then we have a hope for peace, which comes from escaping these desires and rearranging thought within our own minds. It is very much a psychological path of discovery. Primarily, it is to find that sense of reaching Nirvana that absorbs us into the wider universe of life. In short, the great "emptiness" refers to the thought space we require for transcending the "self" and moving toward our goal of achieving Nirvana. In the West, we are overly concerned with the individual person and his or her needs and thirst for self-satisfaction. In the view of Buddha it is necessary to join the life community and to seek a blend that helps us to achieve a more perfect harmony.[30] This seems a pure, practical, and true statement of the ecological perspective. We are no more important than the frog croaking in the pond or the bug biting our arm. We are seeking a becoming that is a blending into a nonmaterial being of all life.

Edward Conze describes the pattern of thought well: "The mind becomes progressively more simple, more renounced, more calm, but it is only for the duration of the Dhyana that this self is forgotten. Wisdom alone can enter the Great Emptiness. It alone can enter the Nirvana which permanently and for all time replaces the impact of sensory stimuli as the force which directs our mind, as long as there is a mind to direct." Later he notes: "The chief purpose of Buddhism is the extinction of separate individuality, which is brought about when we cease to identify anything with ourselves."[31]

"Emptiness" in Buddhism is not an unfilled circle; the term denotes absence of self, not self-effacement. Understanding of "emptiness" comes with the blending of one's separate individuality into that emptiness until there is no distinction between it and the blessed churn of reality. Charles Porter sums Buddha's Noble Truths as: "All living is painful. Suffering is due to craving or desire. Release comes when desire ceases. The way to cessation of suffering is by the Eight-fold Path of 'right views, right intention, right speech, right action, right livelihood, right effort, right mindfulness, right concentration.'"[32] For the social ecologist, this seems an essence of the ultimate environmental sensitivity. Nature does not exist to "serve" us; it exists on its own terms, and we transform ourselves by being absorbed into that nature. Indeed, we are a manifestation of that nature. We become the other as we eat and breathe and become immersed in this life, and in turn it becomes something aware and absorbing of our presence or our belonging as a particle forming this new reality. To harm nature, to treat it as a simple object to serve only our individual desires is to fail as a Buddhist. This attitude is very different from the proselytizing religions—such as Christianity and Islam—which see the earth as a right given for the use and enjoyment of members of its faith. Buddhism as practiced in Bhutan is a more open body of faith that permeates the basic values of the Bhutanese people.

The realities of this faith are important screens of thought that shape the actions of those who see the landscape as something more than an inert object to be exploited for personal gain.[33] However, we should remember that a body of faith does not have absolute authority over the behavior of individual people, it is a guide or a tool for directing us toward a "proper" way. Like devout believers in other faiths, many a devout Buddhist slips past the edicts and guidelines for "appropriate" behavior. So we are talking about central tendencies, not robotic actors.

However, these are central tendencies that vary greatly from those held by foreign advisors with conservation/development dreams. We doubt if the Bhutanese would fully accept the idea that nature is a producer of services primarily directed to serving humanity. Nature exists for itself and operates by its own guidelines.

In Bhutan, the strategy for conservation is to make nonhuman ecosystems a part of school curricula from the early grades through post–high school programs. There is a stress on scientific reality but with a strong underpinning that this is a substantial part of national identity. From the local-level *dzongkhags* to the highest levels of the royal government, the importance of nature is stressed. There is a government agency that serves to remind citizens and policy-makers of the importance of increasing "Gross National Happiness" as a necessary complement to economic matters.[34] The Forestry Department has a community forestry unit, and by 2009 some 250 groups were operating with expectation of continued growth. However, Bhutan's ecosystem conservation is primarily determined through the internalization of a belief system held by individual citizens. It is a psychological rather than an organizational emphasis.

8

Future challenges to the sustainability of human ecosystems are likely to revolve around best practices for encouraging persons and communities to be good stewards of their own habitats. It is only through the cumulative impact of these many decisions that we might have assurance of long-term stewardship. The many large-scale global meetings where top-down messages of country representatives gain some sense of papered-over consensus about environmental problems may be necessary to encourage politicians in the represented countries to "listen up." However, until it infuses local people with a desire and a faith and an enchantment with the wonders of nature, it is unlikely to occur and be maintained. In the long run the ground level is the important thing, for that is where all political and environmental/natural resource issues must ultimately be acted out on a daily basis as well as forever.

This chapter has acknowledged the grand issues of global population growth, climate change, and diminishment of fresh water, and so on. But we chose a menu of three underlying drivers of human ecosystems where HEM concepts could be applied to solving clusters of significant natural resource and environmental issues. The intention here was to analyze historical cases that had characteristics where they served as *ex post facto* experiments and could likely help us lean forward toward better predictions of similar future environmental issues. We considered how property rights institutions joined to norms regulating access to scarce resources have the potential for substantial failure unless the natural resource professional speaks truth to the powerful. We then considered how sustenance institutions linked to socioeconomic resources can be captured by transition from a production to a consumer-driven political economy. In this case the professional is unlikely to stop such trends but must be inventing adaptive modes of response to these new realities. In the third example, the power of environmental forces is expressed in institutions of faith.

Two countries—Nepal and Bhutan—were compared on several dimensions. However, our interest was in how their core faiths—Hinduism and Buddhism—might generate different strategies of conservation action. Nepal has developed a nationwide organizational structure to expand, train, and sustain participatory, community-based systems of resource sustainability. Bhutan has a strategy of psychological empowerment where the person and the local community see their natural setting as an extension of their identity, which is joined and reinforced by the faith of Buddhism and the institutional vision of the royal government. Where Nepal uses external organization forces to sustain their human ecosystem, Bhutan attempts to build it within the person and the community. The test for the natural resource/environmental professional is to envision how these two strategies might be expressed in future research, policy, planning, and management activities.

As with the Amazon region, in the coming decades many predictable human ecosystem problems will emerge for Bhutan, Nepal, and the other high Himalayan countries. In Bhutan, dense urban complexes will arise along the corridor between the cities of Paro and Thimpu in western Bhutan. In Nepal, the rich agricultural and wildlife habitat in the Terai will be converted to urban locales as the rural population moves into the urban ways, but, without a cultural cushion to lean on. Further, the demographics of these places give a strong constraint as the cycle moves from youth to elderly without the resources to provide the benefits that are needed. Thinking how to fit trees, woodlands, and forests into these true mosaics of ecological reality must go beyond a botanical reality to the local social reality, where its purposes may be very different.

Here is the most important reality we need to understand. In Bhutan, Nepal, and many other countries coming into the global web, the reality is that religion is not something simply tacked onto objects of great interest. It is a universe of discourse that contains critical precepts central to guiding human behavior and in balancing the human-nature exchange. Ignoring the importance and reality of religion in matters of social balance and harmony with the ecological realities will result in a plan that is bound to fail. In this sense the tree, woodland, or forest is not the point for starting but the point that is already enriched by the culture of the local people. The tree, woodland, and forest, like us, are part of the larger understanding that transcends its status as a mere object of commerce. The well-meaning stranger needs to come to listen and to learn and might just find that intellectual development goes both ways.

This understanding has been well expressed by Ram Guha: "From an ecological perspective, therefore, peasant movements like Chipko are not merely a defense of the little community and its values, but also an affirmation of a way of life more harmoniously adjusted with natural processes. At one level they are defensive, seeking to escape the tentacles of the commercial economy and the centralizing state; at yet another level they are assertive, actively challenging the ruling-class vision of a homogenizing urban-industrial culture."[35]

T·H·I·R·T·E·E·N

Conclusion

Our brief exploration of the patterns and processes of human ecosystems has been an exercise in trying to make a long story shorter. The complexity of combining human and nonhuman elements is a clear and obvious challenge to our understanding and management of such systems. The usual solution for understanding such complexity is to concentrate observations on the co-variation between a limited set of variables within a limited time-space setting. Important as that approach is, the need to unify these findings and practices may be of even greater importance for meeting future challenges to human ecosystems.

This book shares some of the lessons we have learned in trying to develop and apply a more unifying approach to the study of human ecosystems. It is a learning process approach. We have a template of concepts that helps us to reduce information overload by not attending to all the possible and actual connections in a given system. This conceptual framework and its derived model helps us to identify some of the most crucial linkages in the system and to try out some specific actions that narrow the causal focus.

The HEM has been in use and revision for over forty years. It has adapted to changes over time and will continue to adapt to the changing reality of the problems we hope to understand. Consequently, some of the core variables have been altered, some have been dropped, and others have been added. So far, the major components of the model have remained relatively stable. Many of the case studies we present were done before there was the easily available computing capability to handle Bayesian statistics and complex dynamic models. We hope that future users of the model will adopt potential statistical strategies that are matched to the model's approach.

The utility of the HEM approach is dependent on the active participation of the decision-makers whether they are interested in theoretical research or applied policy, planning, and management. This conceptual model puts the decision-maker and those directly affected by particular environmental/natural resource issues into the center of the decisions. There is no pre-programmed and absolute set of solutions driven by an abstract set of metrics. The experience, the background, and accountability of the people affected by the decision are the drivers of the effort.

Alan Watts, in his 1958 book *Nature, Man, and Woman,* provides a nuanced analysis of Taoism and its relationship to Western science. He emphasizes the moral implications that link thought and practice. He notes:

> For, as another Taoist saying puts it, "When the wrong man uses the right means, the right means work in the wrong way."
>
> Thought, with its serial, one-at-a-time way of looking at things, is ever looking to the future to solve problems which can be handled only in the present—but not in the fragmentary present of fixed and pointed attention. The solution has to be found, as Krishnamurti has said, in the problem and not away from it. And just here, instead of straining toward a future in which one hopes to be different, the mind opens and admits a whole experience in which and by which the problem of what is the "good" of life is answered. In the words of Goethe's *Fragment on Nature*:
>
>> At each moment she starts upon a long, long journey and at each moment reaches her end. . . . All is eternally

present in her, for she knows neither past nor future. For her the present is eternity.¹

So our work circles back. Ours is not a contract with the Devil and a willingness to give up our search for understanding. Rather, we continue Goethe's wager on the resilience of humanity, for we can strive and err and learn and strive on and, as declared in *Faust*, "earn redemption still."²

Notes

Preface

1. See Christopher Marlowe, *The Complete Plays of Christopher Marlowe* (Digireds, 2010); Johann Wolfgang Goethe, *Faust Part Two,* trans. David Luke (1832, repr. ed., Oxford: Oxford University Press, 1994).

ONE Introduction

1. UNFPA (United Nations Population Fund), "Linking Population, Poverty, and Development: Population Trends," *United Nations Population Fund Online,* available at www.unfpa.org/pds/trends.htm, accessed June 6, 2014.

2. Laurie Ann Mazur, "Beyond the Numbers: An Introduction and Overview," in *Beyond the Numbers: A Reader on Population, Consumption, and the Environment,* ed. Laurie Ann Mazur (Washington, DC: Island Press, 1994), 3.

3. Mengpin Ge, Johannes Friedrich, and Thomas Damassa, "6 Graphs Explain the World's Top 10 Emitters," World Resource Institutes (WRI), November 25, 2014, available at www.wri/blog/2014/11/6-graphs-explain-world's-top-10-emitters, accessed October 18, 2016.

4. L. Thurow, "Economic Prosperity Depends Not on Size of Population but on How Fast It Grows," *Technology Review* 89 (1986): 22.

5. Paul R. Ehrlich and Anne Ehrlich, *One with Nineveh: Politics, Consumption, and the Human Future* (Washington, DC: Island Press, 2004), 128.

6. J. G. Speth, *Red Sky at Morning: America and the Crisis of the Global Environment,* 2nd ed. (New Haven, CT: Yale University Press, 2004), 30–33.

7. D. R. Bellwood, T. P. Hughes, C. Foulke, and M. Nystrom, "Confronting the Coral Reef Crisis," *Nature* 429 (2004); L. M. Curran, S. N. Trigg, A. K. McDonald,

D. Astiani, Y. M. Hardiono, P. Siregar, I. Caniago, and E. Kasischke, "Lowland Forest Loss in Protected Areas of Indonesian Borneo," *Science* 303 (2004): 1000; J. A. Thomas, M. G. Telfer, D. B. Roy, C. D. Preston, J. J. D. Greenwood, J. Asher, R. Fox, R. T. Clarke, and J. H. Lawton, "Comparative Losses at British Butterflies, Birds, and Plants and the Global Extinction Crisis," *Science* 303 (2004): 1080.

8. R. M. May, "The Future of Biological Diversity in a Crowded World," *Current Science* 82 (2002): 1330.

9. Gary E. Machlis and Marcia K. McNutt, "Scenario-Building for the Deepwater Horizon Oil Spill," *Science* 239 (2010): 1018.

10. S. W. Mintz, *Sweetness and Power: The Place of Sugar in Modern History* (New York: Penguin Books, 1986), xxvi.

11. Felipe Fernández-Armesto, *Near a Thousand Tables: A History of Food* (New York: Free Press, 2002), xi, 10.

12. Ian Robertson, *Society: A Brief Introduction* (New York: Worth, 1989), 35.

13. For examples of definitions for ecosystem management, see R. Szaro, N. C. Johnson, W. T. Sexton, and A. J. Malik, *Ecological Stewardship: A Common Reference for Ecosystem Management*, vol. 1 (Oxford: Elsevier Science, 1999); M. A. Moote, S. Burke, H. J. Cortner, and M. G. Wallace, *Principles of Ecosystem Management* (Tucson: University of Arizona Press, 1994); S. Leech, A. Wiensczyk, and J. Turner, "Ecosystem Management: A Practitioner's Guide," *BC Journal of Ecosystems and Management* 10 (2009): 2.

14. For example, see C. S. Holling and L. H. Gunderson, "Resilience and Adaptive Cyles," in *Panarchy: Understanding Transformations in Human and Natural Systems*, ed. C. S. Holling and L. H. Gunderson (Washington, DC: Island Press, 2002), 25–62; F. Steiner, *Human Ecology: Following Nature's Lead* (Washington, DC: Island Press, 2002); Millennium Ecosystem Assessment, *Ecosystems and Human Well-Being: Synthesis* (Washington, DC: Island Press, 2005); Richard T. T. Fortman, *Urban Ecology: Science of Cities* (Cambridge, MA: Cambridge University Press, 2014).

15. Peter Heylyn and Henry Seile, *Comosgraphie in Four Books: Containing the Chronographie and Historie of the Whole World, and All the Principall Kingdoms, Provinces, Seas, and Isle Thereof* (London: n.p., 1657), 19.

16. A structuralist approach to human ecology can achieve such insights, if it reveals system functioning that might otherwise go unnoticed. In his essay "The Structuralist Activity," Barthes suggests: "The goal of all structuralist activity, whether reflective or poetic, is to reconstruct [*reconstituer*] an 'object' in such a way as to manifest thereby the rules of its functioning (the 'functions') of this object. Structure is therefore actually a *simulacrum* of the object, but a directed *interested* simulacrum, since the imitated object makes something appear which remained invisible or . . . unintelligible in the natural object." See Barthes, "The Structuralist Activity," in *The Structuralists: From Marx to Levi-Strauss*, ed. R. T. De George, 1st ed. (Garden City, NY: Anchor Books, 1972), 149. While Barthes's comment is framed in literary analysis, its suggestion that an "interested" analysis of structure can provide valuable revelation is also of interest to the human ecologist.

17. S. J. Gould, *The Structure of Evolutionary Theory* (Cambridge, MA: Belknap Press, 2002), 10.

TWO An Overview of the Model

1. E. Callenbach, *Ecology: A Pocket Guide* (Berkeley: University of California Press, 1998), 113.

2. Vaclav Smil, *Energies: An Illustrated Guide to the Biosphere and Civilization* (Cambridge, MA: MIT Press, 1999), 1.

3. For review of energy and social science theory, see E. A. Rosa and Gary E. Machlis, "Energetic Theories of Society: An Evaluation Review," *Sociological Inquiry* 53 (1983); Smil, *Energies*, xiii.

4. Table 2.1 is adapted from Callenbach, *Ecology*, 29.

5. W. R. Catton, *Overshoot: The Ecological Basis of Revolutionary Change* (Urbana: University of Illinois Press, 1982); M. Rees, *Our Final Hour: A Scientist's Warning* (New York: Basic Books, 2003).

6. We describe the components, variables, and flows in more detail in chapter 6.

THREE Lessons and Legacies

1. Rita Reif, *Home: It Takes More than Money* (New York: Quadrangle, 1975), 27.

2. Walter Firey, *Man, Mind, and Land: A Theory* (Glencoe, IL: Free Press, 1960), 226–27.

3. The original wager between Simon and Ehrlich focused on the cost of five metals (copper, chromium, nickel, tin, and tungsten) chosen by Ehrlich in the fall of 1980. Simon believed that the cost of non-government-controlled raw materials would not rise in the long run. The wager was concluded on the original agreed-upon end date in 1990. Simon won the bet, as all five commodities had declined in price during the decade. For a discussion of this wager and others, see Paul Sabin, *The Bet: Paul Ehrlich, Julian Simon, and Our Gamble over Earth's Future* (New Haven: Yale University Press, 2014).

4. E. P. Odum, *Ecology* (New York: Holt, Rinehart, and Winston, 1963).

5. F. H. Bormann and Gene Likens, *Pattern and Processes in a Forested Ecosystem* (New York: Springer-Verlag, 1979).

6. Bormann and Likens, *Pattern and Processes*, 3.

7. Bormann and Likens, *Pattern and Processes*, 5.

8. F. B. Golley, *A History of the Ecosystem Concept in Ecology: More Than the Sum of the Parts* (New Haven, CT: Yale University Press, 1993); F. B. Golley, *A Primer for Environmental Literacy* (New Haven, CT: Yale University Press, 1998).

9. G. E. Likens, F. H. Bormann, R. S. Pierce, J. S. Eaton, and N. M. Johnson, *Biogeochemistry of a Forested Ecosystem* (New York: Springer-Verlag, 1977).

10. C. S. Holling, "Understanding the Complexity of Economic, Ecological, and Social System," *Ecosystems* 4 (2001); J. M. Grove and W. R. Burch Jr., "A Social Ecology

Approach and Applications of Urban Ecosystem and Landscape Analyses: A Case Study of Baltimore, Maryland," *Urban Ecosystems* 1 (1997).

11. Mark Luccarelli, *Lewis Mumford and the Ecological Region: The Politics of Planning* (New York: Guilford Press, 1995); Lewis Mumford, *The Culture of Cities* (1938; repr. New York: Harcourt, Brace, 1970).

12. Robert K. Merton, *On Theoretical Sociology: Five Essays, Old and New* (New York: Free Press, 1967), 66.

13. Jared Diamond, *Guns, Germs, and Steel: The Fates of Human Societies* (New York: W. W. Norton, 1997), 238.

14. Arnold Rose, *Human Behavior and Social Processes* (Boston, MA: Houghton Mifflin, 1962), 3.

15. George H. Mead, *Movements of Thought in the Nineteenth Century*, ed. Merritt H. Moore (Chicago: University of Chicago Press, 1938).

16. Katharine Q. Seelye, "National Parks Proves a Hard Gift to Give," *New York Times*, January 10, 2014, A11.

17. Office of the Press Secretary, "Fact Sheet: President Obama Designates National Monument in Maine's North Woods in Honor of the Centennial of the National Park Service," press release, August 24, 2016, available at www.whitehouse.gov/the-press-office/2016/08/24/fact-sheet-president-obama-designates-national-monument-maines-north.

18. Paul H. Landis, *Three Iron Mining Towns: A Study in Cultural Change* (Ann Arbor, MI: Edwards Brothers, 1938); Harold F. Kaufman and Lois C. Kaufman, *Toward the Stabilization and Enrichment of a Forest Community* (Missoula: University of Montana, 1946).

19. Landis, *Three Iron Mining Towns*, 6–7.

20. Samuel P. Hays, *Conservation and the Gospel of Efficiency: The Progressive Movement, 1890–1920* (Cambridge, MA: Harvard University Press, 1959); Kaufman and Kaufman, *Toward the Stabilization and Enrichment*.

21. Kaufman and Kaufman, *Toward the Stabilization and Enrichment*, 61.

22. Kaufman and Kaufman, *Toward the Stabilization and Enrichment*, 1–2.

23. R. Levins, "Qualitative Analysis of Partially Specified Systems," *Annals of New York Academy of Science* 231 (1974): 123–38; J. H. Milsum, "The Hierarchical Basis for General Living Systems," in *Trends in General Systems Theory*, ed. J. H. Klir (New York: Wiley Interscience, 1972), 145–87; Gordon R. Conway, *Agroecosystem Analysis for Research and Development* (Bangkok: Winrock International, 1986); Howard W. Odum, *Folk, Region, and Society* (Chapel Hill: University of North Carolina Press, 1964); Howard W. Odum and Henry Moore, *American Regionalism: A Cultural-Historical Approach to National Integration* (New York: Henry Holt, 1938); Mumford, *Culture of Cities*.

24. Likens et al., *Biogeochemistry of a Forested Ecosystem*, 1.

25. William R. Burch Jr., Don DeLuca, Gary Machlis, Laurel Burch-Minakan, and Carol Zimmerman, *Handbook for Assessing Energy-Society Relations* (New Haven, CT: Yale University School of Forestry and Environmental Studies, 1977); William R. Burch and Don DeLuca, *Measuring the Social Impact of Natural Resource Policies* (Albuquerque: University of New Mexico Press, 1984).

26. Glenn R. Vernam, *Man on Horseback* (Lincoln: University of Nebraska Press, 1972); Frank Gilbert Roe, *The Indian and the Horse* (1955, repr. Norman: University of Oklahoma Press, 1974).

27. L. Walker and John Pestle, "Air Pollution, Water Pollution and Solid Waste in New Haven, Connecticut: History, Bibliography, Materials and Energy Balance," working paper, Institution of Social and Policy Studies-Yale University (New Haven, CT: Yale University Press, 1971); Leonard Doob, *Sustainers and Sustainability: Attitudes, Attributes, and Actions for Survival* (Westport, CT: Praeger, 1995).

28. T. F. H Allen and T. W. Hoekstra, *Toward a Unified Ecology* (New York: Columbia University Press, 1992), 281.

FOUR The Ecosystem Concept in Biology

1. G. E. Likens and F. H. Bormann, *Biogeochemistry of a Forested Ecosystem*, 2nd ed. (New York: Springer-Verlag, 1995), 1; Golley, *History of the Ecosystem Concept*, 203.

2. Howard W. Odum, *Southern Regions of the United States* (Chapel Hill: University of North Carolina Press, 1936).

3. P. A. Sorokin, *Contemporary Sociological Theories* (New York: Harper & Brothers, 1928), 207.

4. W. G. Sumner and A. G. Keller, *The Science of Society* (New Haven, CT: Yale University Press, 1927); E. Durkheim, *The Division of Labor in Society* (Glencoe, IL: Free Press, 1933); Catton, *Overshoot*; L. Wirth, *The Ghetto* (Chicago: University of Chicago Press, 1928); Firey, *Man, Mind, and Land*; O. D. Duncan, "Social Organization and the Ecosystem," in *Handbook of Modern Sociology*, ed. R. E. L. Farris (Chicago: Rand McNally, 1964), 37–82; P. Selznick, *TVA and the Grass Roots: A Study in the Sociology of Formal Organization* (Berkeley & Los Angeles: University of California Press, 1949); W. F. Ogburn, *Social Change with Respect to Culture and Original Nature* (New York: Viking, 1950); W. F. Cottrell, *Energy and Society: The Relation between Energy, Social Change, and Economic Development* (Westport, CT: Greenwood Press, 1955); K. G. Marx, *Theories of Surplus Value*, vol. 3 (London: Lawrence & Wishart, 1972); M. Weber, *Economy and Society: An Outline of Interpretive Sociology*, ed. G. Roth and C. Wittich (New York: Bedminster Press, 1968); P. C. West, *Natural Resource Bureaucracy and Rural Poverty: A Study in the Political Sociology of Natural Resources* (Ann Arbor: University of Michigan, 1982).

5. J. R. Udry, "Sociology and Biology: What Biology Do Sociologists Need to Know?" *Social Forces* 73 (1995); Allen and Hoekstra, *Toward a Unified Ecology*.

6. A. G. Tansley, "The Use and Abuse of Vegetational Concepts and Terms," *Ecology* 16 (1935).

7. J. Phillips, "Succession, Development, the Climax, and the Complex Organism: An Analysis of Concepts, Part 1," *Journal of Ecology* 22 (1934); Tansley, "The Use and Abuse of Vegetational Concepts and Terms," 284–85, 299.

8. Tansley, "Use and Abuse," 303.

9. Tansley, "Use and Abuse," 306.

10. Robert Edward Cook, "Raymond Lindeman and the Trophic-Dynamic Concept in Ecology," *Science* 198 (1977).

11. R. L. Lindeman, "The Trophic-Dynamic Aspect of Ecology," *Ecology* 23 (1942): 400.

12. Lindeman, "Trophic-Dynamic Aspect," 399–400.

13. Kenneth Boulding, *Principles of Economic Policy* (Englewood Cliffs, NJ: Prentice-Hall, 1958).

14. Gould, *Structure of Evolutionary Theory*, 576.

15. E. P. Odum, *Fundamentals of Ecology*, 2nd ed. (Philadelphia: W. B. Saunders, 1959), 10.

16. Odum, *Fundamentals of Ecology*, 2nd ed., 11.

17. For a cogent description, see Golley, *History of the Ecosystem Concept*.

18. R. Margalef, "On Certain Unifying Principles in Ecology," *American Naturalist* 97 (1963): 357–58.

19. Howard T. Odum, *Systems Ecology: An Introduction* (New York: John Wiley and Sons, 1983), 17.

20. T. S. Kuhn, *The Structure of Scientific Revolutions* (Chicago: University of Chicago Press, 1962); R. V. O'Neill, "Is It Time to Bury the Ecosystem Concept?" *Ecology* 82:12 (2001): 3276.

21. O'Neill, "Is it Time," 3276–3277.

22. O'Neill, "Is it Time," 3277.

23. O'Neill, "Is it Time," 3279.

24. O'Neill, "Is it Time," 3282.

25. O'Neill, "Is it Time," 3282.

FIVE The Roots of Human Ecology

1. D. R. Field and W. R. Burch Jr., *Rural Sociology and the Environment (RSS 50th Anniversary Book Series)* (Westport, CT: Greenwood Press, 1988); M. Micklin, "The Ecological Perspective in the Social Sciences: A Comparative Overview," in *Sociological Human Ecology: Contemporary Issues and Applications*, ed. J. Micklin and H. M. Choldin (Boulder, CO: Westview Press, 1984); R. E. Park and E. W. Burgess, *Introduction to the Science of Sociology* (Chicago: University of Chicago Press, 1921); R. E. Park and R. H. Turner, *Robert E. Park on Social Control and Collective Behavior: Selected Papers* (Chicago: University of Chicago Press, 1967); E. W. Burgess, "Can Neighborhood Work Have a Scientific Basis?," in *The City*, ed. R. E. Park, E. W. Burgess, and R. D. McKenzie (Chicago: University of Chicago Press, 1926), 145; Milla Aïssa Alihan, *Social Ecology: A Critical Analysis* (New York: Columbia University Press, 1938), 9; Lewis Mumford, *Technics and Civilization* (New York: Harcourt, Brace and Company Inc., 1934).

2. E. Gaziano, "Ecology Metaphors as Scientific Boundary Work: Innovation and Authority in Interwar Sociology and Biology," *American Journal of Sociology* 101 (1996).

3. W. I. Thomas and F. Znaniecki, *The Polish Peasant in Europe and America: Monograph of an Immigrant Group*, vol. 3 (Boston, MA: Gorham Press, 1919); H. W.

Zorbaugh, *The Gold Coast and the Slum* (1929, repr. Chicago: University of Chicago Press, 1983); F. M. Thrasher, *The Gang: A Study of 1,313 Gangs in Chicago* (Chicago: University of Chicago Press, 1936); C. R. Shaw, F. M. Zorbaugh, H. D. McKay, and L. S. Cottrell, *Delinquency Areas* (Chicago: University of Chicago Press, 1929); R. E. L. Faris and H. W. Dunham, *Mental Disorders in Urban Areas: An Ecological Study of Schizophrenia and Other Psychoses* (Chicago: University of Chicago Press, 1939); Wirth, *Ghetto*.

4. William Wallace Weaver, "West Philadelphia: A Study of the Natural Social Areas," PhD diss. (University of Pennsylvania, 1930); F. A. Ross, "Ecology and the Statistical Method," *American Journal of Sociology* 38 (1933).

5. Alihan, *Social Ecology*; O. D. Duncan, "From Social System to Ecosystem," in *Urban Patterns: Studies in Human Ecology,* ed. G. A. Theodorson (University Park: Pennsylvania State University Press, 1982).

6. Alihan, *Social Ecology,* 199.

7. Alihan, *Social Ecology,* 246.

8. O. D. Duncan, "Human Ecology and Population Studies," in *The Study of Population: An Inventory and Appraisal,* ed. Philip M. Hauser and O. D. Duncan (Chicago: University of Chicago Press, 1959); Catton, *Overshoot*; Duncan, "Social Organization and the Ecosystem."

9. T. Dietz and E. A. Rosa, "Rethinking the Environmental Impacts of Population, Affluence and Technology," *Human Ecology Review* 1 (1994); Paul R. Ehrlich and A. H. Ehrlich, *Population, Resources, Environment: Issues in Human Ecology* (San Francisco: W. H. Freeman, 1970); Kenneth D. Bailey, "From Poet to Pistol: Reflections on the Ecological Complex," *Sociological Inquiry* 60 (1990).

10. Lee R. Dice, *Man's Nature and Nature's Man: The Ecology of Human Communities* (London: Greenwood Press, 1955), 1.

11. G. D. Pickford and E. H. Reid, "Competition of Elk and Domestic Livestock for Summer Range Forage," *Journal of Wildlife Management* 7 (1943); Dice, *Man's Nature and Nature's Man,* 6–7.

12. Dice, *Man's Nature and Nature's Man,* 251.

13. Dice, *Man's Nature and Nature's Man,* 252.

14. Marston Bates, "Human Ecology," in *Anthropology Today: An Encyclopedic Inventory,* ed. A. L. Kroeber (Chicago: University of Chicago Press, 1953); R. A. Rappaport, *Pigs for the Ancestors: Ritual in the Ecology of a New Guinea People* (New Haven, CT: Yale University Press, 1968).

15. Jonathan Berger and John W. Winton, *Water, Earth, and Fire: Land Use and Environmental Planning in the New Jersey Pine Barrens* (Baltimore: Johns Hopkins University Press, 1985); Frederick Steiner, *The Living Landscape: An Ecological Approach to Landscape Planning* (New York: McGraw Hill, 2000); Frederick Steiner, *Urban Ecological Design: A Process for Regenerative Places* (Washington, DC: Island Press, 2008).

16. J. W. Bennett, *The Ecological Transition: Cultural Anthropology and Human Adaptation* (New York: Pergamon Press, 1976), 52; Yehudi A. Cohen, ed., *Man in Adaptation: The Biosocial Background* (Chicago: Aldine, 1968); Yehudi A. Cohen, ed., *Man in*

Adaptation: The Cultural Present (Chicago: Aldine, 1968); Yehudi A. Cohen, ed., *Man in Adaptation: The Institutional Framework* (Chicago: Aldine, 1971).

17. Roy Ellen, *Environment, Subsistence, and System* (New York: Cambridge University Press, 1982); E. F. Moran, "Ecosystem Ecology in Biology and Anthropology: A Critical Assessment," in *The Ecosystem Approach in Anthropology: From Concept to Practice,* ed. E. F. Moran (Ann Arbor: University of Michigan Press, 1990).

18. W. G. East, *The Geography behind History: No. 419* (New York: W. W. Norton, 1965); B. L. Turner II, W. C. Clark, R. W. Kates, J. F. Richards, J. T. Mathews, and W. B. Meyer, *The Earth as Transformed by Human Action: Global and Regional Changes in the Biosphere over the Past 300 Years* (Cambridge, MA: Cambridge University Press with Clark University, 1990).

19. J. Liu, T. Dietz, S. R. Carpenter, M. Alberti, C. Folke, E. Moran, A. N. Pell, P. Deadman, T. Kratz, J. Lubchenco, E. Ostrom, Z. Ouyang, W. Provencher, C. L. Redman, S. H. Schneider, and W. W. Taylor, "Complexity of Coupled Human and Natural Systems," *Science* 317 (2007); H. E. Kuchka, "The Method for Theory: A Prelude to Human Ecosystems," *Journal of Ecological Anthropology* 5 (2001).

20. G. E. Machlis, J. E. Force, and W. R. Burch, "The Human Ecosystem, Part I: The Human Ecosystem as an Organizing Concept in Ecosystem Management," *Society and Natural Resources* 10 (1997); J. E. Force and G. E. Machlis, "The Human Ecosystem, Part II: Social Indicators for Ecosystem Management," *Society and Natural Resources* 10 (1997).

21. Thomas K. Rudel, "Critical Regions, Ecosystem Management, and Human Ecosystem Research," *Society and Natural Resources* 12 (1999): 258.

22. V. A. Luzadis, K. M. Goslee, E. J. Greenfield, and T. D. Schaeffer, "Toward a More Integrated Ecosystem Model," *Society and Natural Resources* 15 (2002); Jocelyn Forbush, Laura Dunleavy, and Katharine Bill, "The Human Ecosystem Model in Baltimore: Applying a Tool to See Opportunity," term paper from Yale School of Forestry and Environmental Studies, Yale University, 1999.

23. J. M. Hurley, C. Ginger, and D. E. Capen, "Property Concepts, Ecological Thought, and Ecosystem Management: A Case of Conservation Policymaking in Vermont," *Society and Natural Resources* 15 (2002); C. S. Shafer, B. K. Lee, and S. Turner, "A Tale of Three Greenway Trails: User Perceptions Related to Quality of Life," *Landscape and Urban Planning* (2000); E. Stratford and J. Davidson, "Capital Assets and Intercultural Borderlands: Socio-Cultural Challenges for Natural Resource Management," *Journal of Environmental Management* 66 (2002); Frank L. Farmer and Stan L. Albrecht, "The Biophysical Environment and Human Health: Toward Understanding the Reciprocal Effects," *Society and Natural Resources* 11 (1998); J. Parkins, "Enhancing Social Indicators Research in a Forest-Dependent Community," *Forestry Chronicle* 75 (1999); A. Singh, B. Moldan, and T. Loveland, "Making Science for Sustainable Development More Policy Relevant: New Tools for Analysis," *ICSU Series on Science for Sustainable Development* 8 (2002).

24. E. Dellinger and N. Chambers, *Social Indicators for Sonoran Desert Ecosystem Monitoring* (Tucson, AZ: Sonoran Institute, 2004); R. M. Marshall, S. Anderson,

M. Batcher, P. Comer, S. Cornelius, R. Cox, A. Gondor, D. Gori, J. Humke, R. Paredes Aguilar, I. E. Parra, and S. Schwartz, *An Ecological Analysis of Conservation Priorities in the Sonoran Desert Ecoregion,* joint publication of the Nature Conservancy (Arizona Chapter), Sonoran Institute, and El Instituto del Medio Ambiente y el Desarrollo Sustentable del Estado de Sonora, 2000; S. J. Phillips and P. W. Comus, *A Natural History of the Sonoran Desert* (Tucson: Arizona-Sonora Desert Museum Press, 2000).

25. S. T. A. Pickett, W. R. Burch Jr., S. E. Dalton, T. W. Foresman, J. M. Grove, and R. Rowntree, "A Conceptual Framework for the Study of Human Ecosystems in Urban Areas," *Urban Ecosystems* 1997 (1997); Grove and Burch, "Social Ecology Approach"; J. Morgan Grove, Mary L. Cadenasso, Stewart T. A. Pickett, Gary Machlis, and William R. Burch Jr., *The Baltimore School of Urban Ecology: Space, Scale and Time for the Study of Cities* (New Haven, CT: Yale University Press, 2015).

26. P. J. Marcotullio, G. Boyle, S. Ishii, S. K. Karn, K. Suzuki, M. A. Yusuf, and S. Zandaryaa, *Defining an Ecosystem Approach to Urban Management and Policy Development* (United Nations University Institute of Advanced Studies Report, 2003); Mohammad Abu Yusuf, "Integrated Approach of Solid Waste Management to Achieve Urban Sustainability: Asian Experience—Case Study of Bangkok and Dhaka City," vol. 2, Proceedings of the International Symposium on Environmental Management—"Air Pollution and Urban Solid Waste Management and Related Policy Issues," Kanazawa University, Japan, 2004.

27. Machlis and McNutt, "Scenario-Building for the Deepwater Horizon Oil Spill."

six Key Components and Variables for Analyzing Human Ecosystems

1. E. A. Rosa, G. E. Machlis, and K. M. Keating, "Energy and Society," *Annual Review of Sociology* 14 (1988); Cottrell, *Energy and Society,* 2.

2. W. Zelinsky, *The Cultural Geography of the United States* (Englewood Cliffs, NJ: Prentice-Hall, 1973); Turner et al., *The Earth as Transformed by Human Action*; George Galster, *Driving Detroit: The Quest for Respect in the Motor City* (Philadelphia: University of Pennsylvania Press, 2014).

3. M. Reisner, *Cadillac Desert: The American West and Its Disappearing Water* (New York: Penguin Books, 1986).

4. Likens et al., *Biogeochemistry of a Forestry Ecosystem*; G. E. Likens, C. T. Driscoll, and D. C. Busco, "Long-Term Effects of Acid Rain Response and Recovery of a Forestry Ecosystem," *Science* 272 (1996).

5. E. Morales, *Cocaine: White Gold Rush in Peru* (Tucson: University of Arizona Press, 1989).

6. W. B. Clapham, *Human Ecosystems* (New York: Macmillan, 1981).

7. E. O. Wilson, *The Diversity of Life* (New York: Norton, 1992); F. H. Bormann, D. Balmori, and G. T. Geballe, *Redesigning the American Lawn* (New Haven, CT: Yale University Press, 1993).

8. L. von Bertalanffy, *General Systems Theory* (New York: Braziller, 1968); E. O. Wilson, *Sociobiology: The New Synthesis* (Cambridge, MA: Belknap Press of Harvard University Press, 1975); E. O. Wilson, *On Human Nature* (Cambridge, MA: Harvard University Press, 1978); A. H. Hawley, *Human Ecology: The Theory of Community Structure* (New York: Ronald Press, 1950); Burch and DeLuca, *Measuring the Social Impact*; Machlis et al., "Human Ecosystem, Part I."

9. A. H. Hawley, *Human Ecology: A Theoretical Essay* (Chicago: University of Chicago Press, 1986); Durkheim, *Division of Labor in Society*; Turner et al., *The Earth as Transformed by Human Action*; C. Geertz, *Agricultural Involution* (Berkeley: University of California Press, 1963).

10. P. Thompson, *The Nature of Work: An Introduction to Debates on the Labour Process* (London: Macmillan, 1983).

11. R. S. Eckaus, *Basic Economics* (Boston, MA: Little Brown, 1972), 629; C. R. McConnell, *Economics*, 6th ed. (New York: McGraw-Hill, 1975).

12. For more on this, see Mumford, *Technics and Civilization*.

13. E. O. Wilson, *On Human Nature*.

14. G. A. Theodorson and A. G. Theodorson, *Modern Dictionary of Sociology* (New York: Thomas Y. Crowell, 1969); R. Boudon and F. Bourricaud, *A Critical Dictionary of Sociology* (Chicago: University of Chicago Press, 1989); E. Durkheim, *The Rules of Sociological Method*, 8th ed. (New York: Free Press, 1938); R. E. Dunlap, M. E. Kraft, and E. A. Rosa, *Public Reactions to Nuclear Waste: Citizens' Views of Repository Siting* (Durham, NC: Duke University Press, 1993).

15. B. Malinowski, *Magic, Science, and Religion and Other Essays* (Glencoe, IL: Free Press, 1948); W. R. Burch, *Daydreams and Nightmares: A Sociological Essay on the American Environment* (New York: Harper & Row, 1971); D. Worster, *Under Western Skies: Nature and History in the American West* (New York: Oxford University Press, 1992);

16. Karen Armstrong, *A Short History of Myth* (New York: Canon Gate, 2005), 3–4.

17. V. G. Rodwin, *The Health Planning Predicament: France, Quebec, England, and the United States* (Berkeley: University of California Press, 1984).

18. J. Rawls, *A Theory of Justice* (Cambridge, MA: Belknap Press of Harvard University Press, 1971); W. G. Runciman, *Relative Deprivation and Social Justice* (London: Routledge & Kegan Paul, 1966).

19. Durkheim, *Rules of Sociological Method*; M. Weber, *The Protestant Ethic and the Spirit of Capitalism* (London: Allen & Unwin, 1976).

20. Donald J. Hughes, *Ecology in Ancient Civilizations* (Albuquerque: University of New Mexico Press, 1975); E. O. Wilson, *The Creation: An Appeal to Save Life on Earth* (New York: Norton, 2006); Pope Francis, *On Care for Our Common Home: Encyclical Letter Laudato Si' of the Holy Father Francis* (Vatican: Libreria Editrice Vaticana, 2015).

21. Durkheim, *Division of Labor in Society*; G. E. Machlis and W. R. Burch Jr., "Relations Between Strangers: Cycles of Structure and Meaning in Tourist Systems," *Sociological Review* 31 (1983); West, *Natural Resources Bureaucracy and Rural Poverty*.

22. C. E. Bidwell and N. E. Friedkin, "The Sociology of Education," in *Handbook of Sociology*, ed. N. J. Smelser (Newbury Park, CA: Sage, 1988).

23. N. H. Cheek and W. R. Burch Jr., *The Social Organization of Leisure in Human Society* (New York: Harper & Row, 1976); Burch and DeLuca, *Measuring the Social Impact*; J. B. Schor, *The Overworked American: The Unexpected Decline of Leisure* (New York: Basic Books, 1992).

24. K. L. Shell, *The Democratic Political Process* (Waltham, MA: Blaisdell, 1969).

25. Hawley, *Human Ecology*; Field and Burch, *Rural Sociology and the Environment*. For more examples, see Gary E. Machlis, Jo Ellen Force, and Randy Guy Balice, "Timber, Minerals, and Social Change: An Exploratory Test of Two Resource-Dependent Communities," *Rural Sociology* 55 (1990): 411–24; Jo Ellen Force, Gary E. Machlis, and Lianjun Zhang, "The Engines of Change in Resource-Dependent Communities," *Forest Science* 46 (2000): 410–21.

26. See also Gary E. Machlis and T. Hanson, "Warfare Ecology," *Bioscience* 58 (2008): 729–36.

27. Burch and DeLuca, *Measuring the Social Impact*.

28. Burch and DeLuca, *Measuring the Social Impact*.

29. Bormann and Likens, *Pattern and Processes*; Turner et al., *The Earth as Transformed by Human Action*.

30. S. N. Eisenstadt, *From Generation to Generation* (Glencoe, IL: Free Press, 1956); S. Weitz, *Sex Roles: Biological, Psychological, and Social Foundations* (New York: Oxford University Press, 1977); N. S. Abercrombie, S. Hill, and B. S. Turner, *The Penguin Dictionary of Sociology* (New York: Penguin Books, 1988).

31. Abercrombie et al., *Penguin Dictionary of Sociology*.

32. D. H. Wrong, *Power: Its Forms, Bases, and Uses* (Chicago: University of Chicago Press, 1988); M. Scialfa, *An Ethnographic Analysis of Poachers and Poaching in Northern Idaho and Eastern Washington* (Moscow: University of Idaho, 1992).

33. Wrong, *Power*; M. Mann, *The Sources of Social Power: Volume 1: A History of Power from the Beginning to A.D. 1760* (New York: Cambridge University Press, 1984); G. E. Lenski, *Power and Privilege: A Theory of Social Stratification* (Chapel Hill: University of North Carolina Press, 1984); W. J. Goode, *The Celebration of Heroes: Prestige as a Social Control System* (Berkeley: University of California Press, 1978).

34. Reisner, *Cadillac Desert*.

35. Wilson, *Sociobiology*; Torsten Malmberg, *Human Territoriality: Survey of Behavioral Territories in Man with Preliminary Analysis and Discussion of Meaning* (The Hague: Mouton Publishers, 1980).

36. Wilson, *Sociobiology*, 261.

37. Ehrlich and Ehrlich, *One With Nineveh*, 113.

38. Jared Diamond, *Collapse: How Societies Choose to Fail or Succeed* (New York: Viking, 2005).

39. Bennett, *The Ecological Transition*, 31.

40. As we have stressed, the HEM is an evolving instrument of understanding. We have elected to show for each application the HEM as it was actually used in the application, each with its slight to modest variations.

SEVEN Goals, Strategies, and Tactics for Inquiry and Action

1. Christa Sammons, *Goethe the Scientist* (New Haven, CT: Beinecke Rare Book & Manuscript Library/Yale University, 1999).
2. Merton, *On Theoretical Sociology*, 157.
3. W. F. Cottrell, "Death by Dieselization: A Case Study in the Reaction to Technological Change," *American Sociological Review* 16 (1951).
4. Cottrell, "Death by Dieselization," 359.
5. Cottrell, "Death by Dieselization," 360.
6. Cottrell, "Death by Dieselization," 361.
7. Cottrell, "Death by Dieselization," 363.
8. Eric Klinenberg, *Heat Wave: A Social Autopsy of Disaster in Chicago* (Chicago: University of Chicago Press, 2002), 10–11.
9. Klinenberg, *Heat Wave*, 87.
10. Klinenberg, *Heat Wave*, 116.
11. Moran, "Ecosystem Ecology in Biology and Anthropology"; Burch and DeLuca, *Measuring the Social*; Odum, *Systems Ecology*; E. P. Odum, *Ecology and Our Endangered Life-Support Systems* (Sunderland, MA: Sinauer, 1993); K. W. Butzer, "A Human Ecosystem Framework for Archeology," in *The Ecosystem Approach in Anthropology: From Concept to Practice*, ed. E. F. Moran (Ann Arbor: University of Michigan Press, 1990); G. E. Machlis and D. L. Tichnell, *The State of the World's Parks: An International Assessment of Resource Management, Policy, and Research* (Boulder, CO: Westview Press, 1985); G. E. Machlis, "The Contribution of Sociology to Biodiversity Research and Management," *Biological Conservation* 61 (1992); G. E. Machlis, J. E. McKendry, D. J. Forester, and S. T. Engle, *Puget Sound Biodiversity: An Interactive Atlas of Extended Gap Analysis* (Moscow: University of Idaho, 1996); G. E. Machlis and R. G. Wright, "Potential Indicators for Monitoring Biosphere Reserves," in *The Biosphere: Problems and Solutions*, ed. T. N. Vezirohlu (Amsterdam: Elsevier Science, 1984).
12. Mark Sommer, "Beyond the Forest Summit: Activists Strive to Come Up with Ways to Save the Northwest Ancient Trees while Salvaging Area Timber Industry," *Christian Science Monitor*, April 22, 1993; U.S. Department of Agriculture, Forest Service, and U.S. Department of the Interior, Bureau of Land Management, Record of Decision for Amendments to Forest Service and Bureau of Land Management Planning Documents within the Range of the Northern Spotted Owl (Washington, D.C., April 1994).
13. Interior Columbia Basin Ecosystem Management Project (ICBEMP), November 2, 2014, www.icbemp.gov.
14. Force and Machlis, "Human Ecosystem, Part II," 371.
15. Adapted from Force and Machlis, "Human Ecosystem, Part II."
16. M. G. McGown, *The Influence of Organizational Variables on Environmental Management by County Governments*, Unpublished dissertation: University of Idaho, 1994; W. L. Wang Jr., and R. J. Hy, "The Administrative, Fiscal, and Policymaking Capacities of County Governments," *State and Local Law Review* 20 (1988).

17. G. E. Machlis, J. E. Force, and J. E. McKendry, *An Atlas of Social Indicators for the Upper Columbia River Basin* (Moscow: University of Idaho, 1995).

18. ICBEMP, www.icbemp.gov/html/icbhome.html.

EIGHT Using the Model for Science during Crisis

1. Portions of this chapter are adapted from two Operational Group Sandy Technical Progress Reports: Department of the Interior, DOI Strategic Sciences Working Group, *Mississippi Canyon 252/ Deepwater Horizon Oil Spill Progress Report* (Washington, DC, June 2010); Department of the Interior, DOI Strategic Sciences Working Group, *Mississippi Canyon 252/ Deepwater Horizon Oil Spill Progress Report 2* (Washington, DC, April 2012).

2. Report available at www.usgs.gov/oilspill/ docs/SSWG_Progress_Report _09june10.pdf: Department of the Interior, *Mississippi Canyon 252/Deepwater Horizon Oil Spill Progress Report*; Machlis and McNutt, "Scenario-Building for the Deepwater Horizon Oil Spill."

3. Machlis and McNutt, "Scenario-Building for the Deepwater Horizon Oil Spill"; J. Liu et al., "Complexity of Coupled Human and Natural System"; C. S. Holling and L. H. Gunderson, eds., *Panarchy: Understanding Transformations in Human and Natural Systems* (Washington, DC: Island Press, 2002).

4. See J. E. Haas, R. W. Kates, and M. J. Bowden, *Reconstruction Following Disaster* (Cambridge, MA: MIT Press, 1977); R. W. Kates, C. E. Colten, S. Laska, and S. P. Leatherman, "Reconstruction of New Orleans after Hurricane Katrina: A Research Perspective," *Proceedings of the National Academy of Sciences* 103 (2006); Machlis and McNutt, "Scenario-Building for the Deepwater Horizon Oil Spill."

5. D. Burley, P. Jenkins, S. Laska, and T. Davis, "Place Attachment and Environmental Change in Coastal Louisiana," *Organization and Environment* 20 (2007); S. A. Castillo and P. Moreno-Casasola, "Coastal Sand Dune Vegetation: An Extreme Case of Species Invasion," *Journal of Coastal Conservation* 2 (1996); N. N. Rabalais, R. E. Turner, and W. J. Wiseman Jr., "Hypoxia in the Gulf of Mexico," *Journal of Environmental Quality* 30 (2001); J. Tibbetts, "The State of the Oceans, Part I: Eating Away at a Global Food Source," *Environmental Health Perspective* 1125 (2004).

6. C. S. Holling, "Resilience and Stability of Ecological Systems," *Annual Review of Ecology and Systematics* 4 (1973); W. N. Adger, T. P. Hughes, C. Folke, S. R. Carpenter, and J. Rockstrom, "Social-Ecological Resilience to Coastal Disasters," *Science* 308 (2005).

7. Vertical life zones adapted from B. H. Robinson, "Conservation of Deep Pelagic Biodiversity," *Conservation Biology* 23 (2009); major ecosystem types adapted from J. J. McGuire, "Fisheries Topics: Ecosystems, Types of Ecosystems," FAO Fisheries and Aquaculture Department (2005), available at www.fao.org/fishery/topic/3320/en; sociopolitical and administrative units from the local village to parish, county, and state adapted from E. S. Sheppard and R. McMaster, *Scale and Geographic Inquiry: Nature, Society, and Method* (Malden, MA: Blackwell, 2004); D. L. Felder and D. K. Camp, eds.,

Gulf of Mexico Origin, Water, and Biota: Volume 1: Biodiversity (College Station: Texas A & M University Press, 2009).

8. SmartDraw (Hemera Technologies, Inc. San Diego, CA). Any use of trade, product, or firm names is for descriptive purposes only and does not imply endorsement by the U.S. government.

9. C. Weiss, "Expressing Scientific Uncertainty," *Law, Probability, and Risk* 2 (2003).

10. Machlis and McNutt, "Scenario-Building for the Deepwater Horizon Oil Spill"; G. D. Peterson, G. S. Cumming, and S. R. Carpenter, "Scenario Planning: A Tool For Conservation In an Uncertain World," *Conservation Biology* 17 (2003).

11. National Oceanic and Atmospheric Administration, "NOAA's State of the Coast National Coastal Population Report: Population Trends from 1970–2020," *National Coastal Population Report* 22 (2013); E. S. Blake, T. B. Kimberlain, R. J. Berg, J. P. Cangialosi, and J. L. Beven II, "Tropical Cyclone Report Hurricane Sandy 22–29 October 2012: Miami," *National Hurricane Center* 157 (2013); "Response to Hurricane Sandy," *U.S. Geological Survey*, March 21, 2013, http://coastal.er.usgs.gov/hazard-events/sandy; "National Catastrophes and Man-Made Disasters in 2012: A Year of Extreme Weather Events in the US," *SwissRe*, May 1, 2013, http://media.swissre.com/documents/sigma_22013_EN.pdf.

12. Department of the Interior, *Mississippi Canyon 252/Deepwater Horizon Oil Spill Progress Report*; Machlis and McNutt, "Scenario-Building for the Deepwater Horizon Oil Spill"; Department of the Interior, *Mississippi Canyon 252/Deepwater Horizon Oil Spill Progress Report 2*.

NINE Revitalizing Human Communities and Reclaiming Biological Communities

1. Various tables in the U.S. Census Bureau reports were consulted for the demographic data about the city of Baltimore, over the period of interest for this Baltimore Case study.

2. Jane Jacobs, *The Death and Life of Great American Cities* (New York: Vintage, 1992).

3. William R. Burch and Morgan Grove, "People, Trees, and Participation on the Urban Frontier," *Unasylva* 44 (1993); Grove and Burch, "Social Ecology Approach"; William R. Burch, "Challenges and Possible Futures for the Forestry Profession in a Global, Post-Industrial Social Economy: Lessons from Britain," *Scottish Forestry* 56 (2002).

4. Pickett et al., "Conceptual Framework"; Machlis et al., "Human Ecosystem, Part I"; Burch and DeLuca, *Measuring the Social Impact*; Rusong Wang and Zhiyun Ouyang, "A Human Ecology Model for the Tianjin Urban Ecosystem: Integrating Human Ecology, Ecosystem Science, and Philosophical Views into an Urban Eco-Complex Study," in *Understanding Urban Ecosystems*, ed. Alan R. Berkowitz, Charles H. Nilon, and Karen S. Hollweg (New York: Springer, 2003); R. S. Wang, J. Z. Zhao, and X. L. Dai, *Human Ecology in China* (Beijing: China Science and Technology Press, 1989); D. J. Rapport and A. Singh, "Framework for the State of Environment Report," discussion paper, UNEP-SCOPE workshop, "Making Science More Policy Relevant."

5. Likens et al., *Biogeochemistry of a Forested Ecosystem*; F. H. Bormann and G. Likens, *Pattern and Processes*; Golley, *History of the Ecosystem Concept*.

6. Lisa Hite, Mary Porter, Gennady Schwartz, Morgan Grove, Chris Rogers, and Lisa Vernegaard, *Strategic Plan for Action, Baltimore Recreation and Parks* (Baltimore, MD: Department of Recreation and Parks, 1991).

7. Jennifer Aley and William Burch, *Fort McHenry National Monument School Field Trip and Survey Report* (New Haven, CT: URI, 1994); Jennifer Aley and William Burch, *A Comparative Study of Variations in Perceptions and Behaviors of National Park Unit Visitors in the Baltimore Region* (New Haven, CT: URI, 1995); Jennifer Aley and William Burch, *Baltimore Park Visitor Demographics: A Survey of Park Visitors from the Baltimore Region* (New Haven, CT: URI, 1997); Hite et al., *Strategic Plan for Action*; Shawn E. Dalton, William Burch, J. Morgan Grove, and Jennifer Aley, "Fort McHenry National Monument and Historic Shrine Market Study," *Working Paper 8* (New Haven, CT: Yale Urban Resource Initiative, 1992); Shawn E. Dalton, *Protecting Wetlands in Baltimore: Series of Nine Training Manuals and Citizen Guidelines* (Baltimore, MD: Revitalizing Baltimore, Parks and People Foundation, 1993); Shawn E. Dalton, Beth Conover, Dexter Mead, LaTonya Danzy, Yale URI, Marlyn J. Peritt, Lisa Hite, and Mary Porter, *Natural Resource Managers Training Manuel and Field Guide* (Baltimore, MD: Department of Recreation and Parks, 1993); William Burch and Paul Jahnige, *Gwynns Falls Greenway Human Ecology Inventory*, 2 vols. (Baltimore, MD: Parks and People Foundation, 1995); Paul Jahnige, comp., *The Gwynns Falls Watershed Ecological Resource Atlas* (Baltimore, MD: Revitalizing Baltimore, Parks and People Foundation, 1999).

8. Hite et al., *Strategic Plan for Action*.

9. Guy W. Hagar, *Urban and Community Forestry* (grant narrative statement) (New Brunswick, NJ: Rutgers University Press, 2004), 18.

10. "Annual Report: A Year to Reimagine, Restore, and Protect Baltimore's Waterways," Blue Water Baltimore, 2013, 3, http://bluewaterbaltimore.org/about/annual-report-and-financials/.

TEN Toward a More Perfect Civic Order

1. S. Gordon, J. Greenfield, C. Laporte, M. Lowenstein, and C. Rodstrom, "Baltimore Project Final Report," *Working Paper 2* (New Haven, CT: Yale Urban Resource Initiative, 1990).

2. The Urban Resources Initiative (URI) began in Baltimore, Maryland, in 1989 when Burch's students in the Yale Forestry and Environmental Studies program went to Baltimore to apply the principles of social forestry as they worked with neighborhood groups and the Baltimore Department of Parks and Recreation. The first available report from the students' work (with a preface from Burch) was produced in 1990: S. Gordon, J. Greenfield, C. Laporte, M. Lowenstein, and C. Rodstrom, "Working Paper #2, Baltimore Project Final Report," Urban Resources Initiative, School of Forestry and Environmental Studies, Yale University, 1990. This "working paper" is a set of separate papers prepared by the listed authors on various projects undertaken in the neighborhoods of Baltimore. This report along with nearly 50 others over the past two

decades can be found online at the following address: http://environment.yale.edu/uri/publications/. The Urban Resources Initiative was formally incorporated in 1991 and became a stand-alone fiscal and legal nongovernmental organization with a local board of directors. As stated on their website: "Though closely affiliated with the Yale School of Forestry and Environmental Studies, this autonomy assures a governing body whose diverse interests represent all corners of New Haven."

3. Gordon et al., "Working Paper #2, Baltimore Project Final Report."

4. Burch and Jahnige, *Gwynns Falls Greenway Human Ecology Inventory.*

5. Jahnige, *Gwynns Falls Watershed Ecological Resource Atlas.*

6. Aley and Burch, *Comparative Study.*

7. Forbush et al., "The Human Ecosystem Model in Baltimore," i.

8. The current and past vital signs data are available on the BNIA's website at http://bniajfi.org/vital_signs/.

9. Machlis et al., "The Human Ecosystem"; Baltimore Ecosystem Study, "Human Ecological System as a Data Integrator and Interface," www.beslter.org/frame4-page_2_1.html.

10. Pickett et al., "Conceptual Framework," 195–96.

ELEVEN Extending the Capability of the Model

1. William Cronon, *Changes in the Land: Indians, Colonists, and the Ecology of New England* (New York: Hill & Wang, 1983), 10; Sim Van der Ryn and Peter Calthorpe, *Sustainable Communities: A New Design Synthesis for Cities, Suburbs, and Towns* (San Francisco: Sierra Club Books, 1986); John P. Kretzmann and John L. McKnight, "Capturing Local Institutions for Community Building," in *Building Communities from the Inside Out: A Path toward Finding and Mobilizing a Community's Assets* (Skokie, IL: ACTA Publications, 1993), 170–275; Y. S. Rao, Marilyn W. Hoskins, Napoleon T. Vergara, and Charles Castro, *Community Forestry: Lessons from Case Studies in Asia and the Pacific Region* (Bangkok: Regional Office for Asia and the Pacific, Food and Agriculture Organization, UN, 1985).

2. The characteristics of patch analysis have been well developed by the long-term Baltimore Ecosystem Study as described in a book that details how patch analysis combines biological and social science approaches for understanding urban ecosystems. On the Baltimore Ecosystem Study, see Grove et al., *Baltimore School of Urban Ecology*; Stewart T. A. Pickett and Mary L. Cadenasso, "Meaning, Model, and Metaphor of Patch Dynamics," in *Designing Patch Dynamics*, ed. Brian McGrath, Victoria Marshall, M. L. Cadenasso, J. Morgan Grove, S. T. A. Pickett, Richard Plunz, and Joel Towers (New York: Columbia Graduate School of Architecture, 2007), 17–18.

3. Pickett and Cadenasso, "Meaning, Model, and Metaphor."

4. B. T. Bormann, B. K. Williams, G. S. Stankey, and T. Minkova, "Learning to Learn: The Best Available Science of Adaptive Management," unpublished manuscript, 2015; Byron K. Williams and Eleanor D. Brown, *Adaptive Management: U.S. Department of the Interior-Applications Guide* (Washington, DC: U.S. Department of the Interior, 2012).

5. Chris Maser, Bernard T. Bormann, Martha H. Brookes, A. Ross Kiester, and James F. Weigand, "Sustainable Forestry through Adaptive Ecosystem Management Is An Open-Ended Experiment," in *Sustainable Forestry: Philosophy, Science, and Economics* (Delray Beach, FL: St. Lucia Press, 1994), 303–40.

6. G. Evelyn Hutchinson, "Interesting Ways of Thinking of Death," *An Introduction to Population Ecology* (New Haven, CT: Yale University Press, 1980), 41–89; Austin Troy, *The Very Hungry City: Urban Energy, Efficiency, and the Economic Fate of Cities* (New Haven, CT: Yale University Press, 2012).

7. Rebecca Solnit, *A Paradise Built in Hell: The Extraordinary Communities That Arise in Disaster* (New York: Penguin, 2010).

8. S. T. A. Pickett, M. L. Cadenasso, and Brian McGrath, *Resilience in Ecology and Urban Design* (New York: Springer, 2013).

9. Gideon Sjoberg, *The Preindustrial City: Past and Present* (New York: Free Press, 1960); Anthony P. Andrews, *First Cities* (Washington, DC: St Remy Press, 1995); William Skinner, ed., *The City in Late Imperial China* (Taipei: SMC Publishing, 1977); E. Shevky and W. Bell, *Social Area Analysis: Theory, Illustrative Applications, and Computational Procedure* (Stanford, CA: Stanford University Press, 1955); Mumford, *Culture of Cities*; D. Timms, *The Urban Mosaic: Towards a Theory of Residential Differentiation* (Cambridge, MA: Cambridge University Press, 1971), 1; M. J. Weiss, *The Clustering of America* (New York: Tilden Press, 1988).

10. David Ley, *A Social Geography of the City* (London: Harper and Row, 1983), 71.

11. E. W. Burgess, "Can Neighborhood Work Have a Scientific Basis?" in *The City*, ed. R. E. Park, E. W. Burgess, and R. D. McKenzie (Chicago: University of Chicago Press, 1925); Homer Hoyt, *The Structure and Growth of Residential Neighborhoods in American Cities* (Washington, DC: Federal Housing Administration, 1939); Shevky and Bell, *Social Area Analysis*; Ley, *Social Geography of the City*, 84; Walter Firey, "Sentiment and Symbolism as Ecological Variables," *American Sociological Review* (1945): 140–48.

12. William H. Michelson, *Man and His Urban Environment: A Sociological Approach* (New York: Addison-Wesley, 1970), 16–18.

13. Cronon, *Changes in the Land*, 13.

14. For example, see Walter Firey, *Land Use in Central Boston* (Cambridge: Harvard University Press, 1947).

15. William Kornblum, *Blue Collar Community* (Chicago: University of Chicago Press, 1974); Firey, *Land Use in Central Boston*; Pickett and Cadenasso, "Meaning, Model, and Metaphor"; Klinenberg, *Heat Wave*.

16. Kornblum, *Blue Collar Community*, 15.

17. Kornblum, *Blue Collar Community*, 228.

18. Douglas Rae, *Urbanism and Its End* (New Haven, CT: Yale University Press, 2003).

19. Mark Abrahamson, *Urban Enclaves: Identity and Place in America* (New York: St. Martin's Place, 1996); Harold A. McDougall, *Black Baltimore: A New Theory of Community* (Philadelphia: Temple University Press, 1993), 113.

20. Klinenberg, *Heat Wave*.

21. Robert J. Sampson, Stephen W. Raudenbush, and Felton Earls, "Neighborhoods and Violent Crime: A Multilevel Study of Collective Efficacy," *Science* 277 (1997).

22. Klinenberg, *Heat Wave*, 127.

23. Durkheim, *Division of Labor in Society*; Gianfranco Poggi, *Durkheim* (Oxford: Oxford University Press, 2000), 51–52.

24. Jacobs, *Death and Life of Great American Cities*. Steven Johnson, *Emergence: The Connected Lives of Ants, Brains, Cities, and Software* (New York: Scribner, 2001).

25. Sampson et al., "Neighborhoods and Violent Crime."

26. C. McCord and H. P. Freeman, *Excess Mortality in Harlem* (New York: New England Journal of Medicine, 1990); Arline Geronimus, John Bound, Timothy Waidman, Marianne Hillemeier, and Patricia Burns, "Excess Mortality among Blacks and Whites in the United States," *New England Journal of Medicine* 355 (1996); Sherry Olson, *Baltimore: The Building of an American City* (Baltimore, MD: Johns Hopkins University Press, 1997).

27. Poggi, *Durkheim*, 56.

28. M. J. Weiss, *The Clustered World: How We Live, What We Buy, What It All Means About Who We Are* (New York: Little Brown and Company, 2000); J. Morgan Grove, Mary L. Cadenasso, William R. Burch, Jr., Stewart A. Pickett, Kirsten Schwarz, Jarlath O'Neil-Dunne, Mathew Wilson, Austin Troy, and Christopher Boone, "Data and Methods Comparing Social Structure and Vegetation Structure of Urban Neighborhoods in Baltimore, Maryland," *Society and Natural Resources* 19 (2006): 119; C. A. Martin, P. S. Warren, and A. Kinzig, "Neighborhood Socioeconomic Status Is a Useful Predictor of Perennial Landscape Vegetation in Small Parks Surrounding Residential Neighborhoods in Phoenix, Arizona," *Landscape Urban Plan* 69 (2004). The Potential Rating Index for zip code markets was developed by Jonathan Robin in 1974, as described in Claritas, *PRIZM Cluster Snapshots: Getting to Know the 62 Clusters* (Ithaca, NY: Claritas Corporation, 1999).

29. Robert Ardrey, *The Territorial Imperative: A Personal Inquiry into the Animal Origins of Property and Nations* (London: Collins, 1966); Konrad Lorenz, *Evolution and Modification of Behavior* (Chicago: University of Chicago Press, 1973); Lionel Tiger, *Men in Groups* (London: Nelson, 1969); Lionel Tiger and Robin Fox, *The Imperial Animal* (New York: Holt Rinehart & Winston, 1971); N. Tinbergen, "On War and Peace in Animals and Men," *Science* 160 (1968); Wilson, *Sociobiology*; Carl N. Degler, *In Search of Human Nature: The Decline and Revival of Darwinism in American Social Thought* (New York: Oxford University Press, 1991), 225.

30. Wilson, *Sociobiology*, 260.

31. Wilson, *Sociobiology*, 274; Misia Landau, *Narratives of Human Evolution* (New Haven, CT: Yale University Press, 1991), 177.

32. Herbert J. Gans, *The Urban Villagers* (New York: Free Press, 1962).

33. Erving Goffman, *Relations in Public: Microstudies of the Public Order* (New Brunswick, NJ: Transactions Publishers, 2010), 29.

34. Edward T. Hall, *The Silent Language* (Garden City: Doubleday & Co, 1961); Anna Bramwell, *Ecology in the 20th Century: A History* (New Haven, CT: Yale University Press, 1989).

35. Malmberg, *Human Territoriality*, 216.

36. Sampson et al., "Neighborhoods and Violent Crimes," 918–19.

37. Tansley, "Use and Abuse"; Bormann and Likens, *Pattern and Processes*; Golley, *History of the Ecosystem Concept*; Holling, "Understanding the Complexity."

38. J. Morgan Grove, Mary L. Cadenasso, William R. Burch Jr., Stewart A. Pickett, Kirsten Schwarz, Jarlath O'Neil-Dunne, Mathew Wilson, Austin Troy, and Christopher Boone, "Data and Methods Comparing Social Structure and Vegetation Structure of Urban Neighborhoods in Baltimore, Maryland," *Society and Natural Resources* 19 (2006); J. Morgan Grove, A. R. Troy, J. P. M. O'Neil-Dunne, W. R. Bruch, Jr., M. L. Cadenasso, and S. T. A. Pickett, "Characterization of Households and Its Implications for the Vegetation of Urban Ecosystems," *Ecosystems* 9 (2006); Gordon G. Whitney and Stanley D. Adams, "Men as a Maker of New Plant Communities," *Journal of Applied Ecology* 17 (1980): 442; W. Lloyd Warner, *Yankee City*, vol. 1, abridged edition (New Haven, CT: Yale University Press, 1963); W. E. Darrenbacher, *Plants and Landscapes: An Analysis of Ornamental Plantings in Four Berkeley Neighborhoods* (Berkeley: University of California, 1969).

39. Grove et al., "Characterization of Households"; R. K. Merton and A. Kitt, "Contributions to the Theory of Reference Group Behavior," in *Continuities in Social Research: Studies in the Scope and Method of the American Soldier*, ed. R. K. Merton and P. Lazarsfeld (Glencoe, IL: Free Press, 1950); Tomotsu Shibutani, "Reference Groups and Social Control," in *Human Behavior and Social Processes: An Interactionist Approach*, ed. Arnold Rose (Boston, MA: Houghton Mifflin. Co, 1962).

40. Olson, *Baltimore*, xiii.

41. Olson, *Baltimore*, 82.

42. Olson, *Baltimore*, 162.

43. Olson, *Baltimore*, 325.

44. Bernard T. Bormann and A. Ross Kiester, "Options Forestry Acting on Uncertainty," *Journal of Forestry* June (2004): 22.

45. National Environmental Policy Act (NEPA) documents associated with the Five Rivers Project on the Siuslaw National Forest in Oregon. Available at http://data.ecosystemmanagement.org/nepaweb_project_exp.php?project=32675, accessed June 4, 2016.

46. John A. Parrotta and Ronald L. Trosper, *Traditional Forest-Related Knowledge: Sustaining Communities, Ecosystems, and Biocultural Diversity* (New York: Springer, 2012).

47. Pickett and Cadenasso, *Designing Patch Dynamics*, 16–29; Grove et al., *Baltimore School of Urban Ecology*.

48. Pickett and Cadenasso, *Designing Patch Dynamics*, 106–7.

TWELVE Leaning Forward

1. Patricia Cohen, "Study Finds Global Wealth Is Flowing to the Richest," *New York Times*, January 19, 2015: B6; Thomas Piketty, *Capital in the Twenty-First Century*

(Cambridge: Harvard University Press, 2014); Simon Reid-Henry, *The Political Origins of Inequality: Why a More Equal World Is Better for Us All* (Chicago: Chicago University Press, 2015); Joseph Stiglitz, *The Price of Inequality: How Today's Divided Society Endangers the Future* (New York: W. W. Norton, 2012); Richard Wilkinson and Kate Pickett, *The Spirit Level: Why Equality Is Better for Everyone* (London: Penguin Books, 2009).

2. Joseph Epstein, *Snobbery: The American Version* (Boston: Houghton-Mifflin Company, 2002); James B. Twitchell, *Living It Up: America's Love Affair with Luxury* (New York: Simon & Schuster, 2003); Susan J. Matt, *Keeping Up with the Joneses: Envy in American Consumer Society, 1890–1930* (Philadelphia: University of Pennsylvania Press, 2003); David Harvey, *The Urban Experience* (Baltimore, MD: Johns Hopkins University Press, 1989); E. Digby Baltzell, *Philadelphia Gentlemen: The Making of a National Upper Class* (New Brunswick, NJ: Transaction Publishers, 1995); Lenski, *Power and Privilege*; Wilkinson and Pickett, *The Spirit Level*; James C. Scott, *Seeing Like a State: How Certain Schemes to Improve the Human Condition Have Failed* (New Haven, CT: Yale University Press, 1998).

3. Joe Jackson, *The Thief at the End of the World: Rubber, Power, and the Seeds of Empire* (New York: Penguin Books, 2008).

4. Greg Grandin, *Fordlandia: The Rise and Fall of Henry Ford's Forgotten Jungle City* (New York: Picador, 2009).

5. Jackson, *Thief at the End of the World*, 302.

6. "Policy Analysis Report," City and County of San Francisco Board of Supervisors, Budget and Legislative Analyst's Office, February 10, 2014.

7. Samuel Butler, *Erewhon* (New York: Lancer Books, 1872); David Streitfeld, "Still No. 1, and Doing What He Wants," *New York Times*, April 12, 2014; Jon Mooallem, "Larry Ellison Bought an Island in Hawaii. Now What?" *New York Times Magazine*, September 23, 2014.

8. Stephen Ponder, "Iowa in the Amazon," *New York Times*, November 25, 2013, A23.

9. John Berger, "A Jerome of Photography: The Camera as an Instrument of Knowledge," *Harper's Magazine*, December 2005, 87.

10. Alan Durning, *How Much Is Enough? The Consumer Society and the Future of the Earth* (New York: W. W. Norton, 1992); Ramachandra Guha, *The Unquiet Woods: Ecological Change and Peasant Resistance in the Himalaya* (Berkeley: University of California Press, 2000); Jerome O. Steffen, *The Tragedy of Abundance: Myth Restoration in American Culture* (Boulder: University Press of Colorado, 1993); Robert Frank, *Richistan: A Journey through the American Wealth Boom and the Lives of the New Rich* (New York: Three Rivers Press, 2007).

11. Bill Bishop, *The Big Sort* (Boston, MA: Houghton-Mifflin Harcourt, 2008); Thomas Hine, *I Want That! How We All Become Shoppers* (New York: Perennial/Harper & Row, 2003).

12. Tim Kasser, *The High Price of Materialism* (Cambridge, MA: MIT Press, 2002); David G. Myers, *The American Paradox: Spiritual Hunger in an Age of Plenty* (New Haven, CT: Yale University Press, 2001); William Rathje and Cullen Murphy, *Rubbish! The*

Archaeology of Garbage, What Our Garbage Tells Us About Ourselves (New York: Harper Collins, 1992).

13. Walter Prescott Webb, *The Great Frontier* (Austin: University of Texas Press, 1964), 94; Steffen, *Tragedy of Abundance*, 7–9.

14. Brian Czech, *Shoveling Fuel for a Runaway Train: Errant Economists, Shameful Spenders, and a Plan to Stop Them All* (Berkeley: University of California Press, 2000); Simon Schama, *Landscape and Memory* (New York: Vintage Books, 1995); Clarence Glacken, *Traces on the Rhodian Shore: Nature and Culture in Western Thought from Ancient Times to the End of the Eighteenth Century* (Berkeley: University of California Press, 1976).

15. Max Weber, *The Protestant Ethic and the Spirit of Capitalism* (New York: Scribner, 1958).

16. Weber, *Protestant Ethic and the Spirit of Capitalism*.

17. William Easterly, *The White Man's Burden* (New York: Penguin Books, 2007); Jack D. Ives and Bruno Messerli, *The Himalayan Dilemma: Resolving Development and Conservation* (New York: Routledge, 1989); Benjamin R. Barber, *Jihad vs. McWorld: How Globalism and Tribalism Are Reshaping the World* (New York: Ballantine Books, 1995).

18. Federation of Community Forestry Users Nepal (FECOFUN), www.fecofun.org.np, accessed November 23, 2014; William R. Burch, "Lessons from the High Himalayas for Natural Resource Conservation," in *International Conference on Alishan Centary Forestry*, ed. H. J. Wang (Chiayi, Taiwan: n.p., 2011).

19. Ives and Messerli, *Himalayan Dilemma*; Narayan Khadka, *Foreign Aid, Poverty, and Stagnation in Nepal* (New Delhi: Vikas Publishing, 1991); Reuben Ellis, *Vertical Margins, Mountaineering, and the Landscape of Neoimperialism* (Madison: University of Wisconsin Press, 2001); Pankaj Mishra, *Temptations of the West* (New York: Picador Press, 2007); Robert S. Anderson and Walter Huber, *The Hour of the Fox: Tropical Forests, the World Bank, and Indigenous People in Central India* (Seattle: University of Washington Press, 1988).

20. Khadka, *Foreign Aid, Poverty, and Stagnation*; P. P. Karan and Hiroshi Ishii, *Nepal: Development and Change in a Landlocked Himalayan Kingdom* (Tokyo: University of Foreign Studies, 1994); Dor Bahadur Bista, *Fatalism and Development* (Calcutta: Orient Longman, 1991); Easterly, *White Man's Burden*.

21. Jonathan Kusel and Elisa Adler, *Forest Communities, Community Forests* (Lanham, MD: Rowman & Department of Forestry Services, Royal Government of Bhutan, 2003).

22. Trilok Chandra Majupuria, *Religion and Useful Plants of Nepal and India* (Bangkok: Craftsman Press, 1988); Rajesh Gautam and Thapa-Magar Asokek, *Tribal Ethnography of Nepal*, vols. 1 and 2 (Delhi: Book Faith India, 1994).

23. Zangley Dupka, *Bhutan and Its Natural Resources* (New Delhi: Vikas Publishing House, 1991).

24. J. Baird Callicott and Roger T. Ames, *Nature in Asian Traditions of Thought: Essays in Environmental Philosophy* (Albany: State University of New York Press, 1989); Hajime Nakamura and Philip P. Wiener, *Ways of Thinking of Eastern Peoples: India, China, Tibet, Japan* (Albany: State University of New York Press, 1981).

25. Kenneth Morgan, *The Religion of the Hindus* (Delhi: Shiri Jainendra Press, 1987).

26. Jagannath Adhikari, "Hindu Tradition and Peasant Farming in the Foothills of Nepal," in *Sustainable Agriculture and the World's Religious Traditions,* ed. Todd LeVasseur, Pramod Parajuli, and Norman Wirzba (Lexington: University Press of Kentucky, 2015), 1.

27. E. F. Schumacher, *Small Is Beautiful* (New York: Harper & Row, 1973); Adhikari, "Hindu Tradition and Peasant Farming," 19.

28. Federation of Community Forestry Users Nepal, 2010.

29. Federation of Community Forestry Users Nepal, 2013.

30. The Karmapa, Ogyen Trinley Dorje, *The Heart Is Noble: Changing the World from the Inside Out* (Boston, MA: Shambhala South Asian Editions, 2013).

31. Edward Conze, *Buddhism: Its Essence and Development* (New Delhi: Munshiram Manoharlal, 1994), 101–7.

32. Charles Francis Porter, *The Great Religious Leaders* (New York: Washington Square Press, 1962), 134.

33. The Karmapa, *The Heart Is Noble.*

34. Jigme Singye Wangchuck, who became Bhutan's king in 1972, formulated "Gross National Happiness" as a quality of life indicator in Bhutan. It represents an "ethos of environmental sustainability, cultural preservation, and "holistic" civic contentment. Jody Rosen, "Bhutan: A Higher State of Being," *New York Times,* October 30, 2014, available at www.nytimes.com/2014/10/30/t-magazine/Bhutan-bicycle-gross-national-happiness.html, accessed October 19, 2016.

35. Ramachandra Guha, *The Unquiet Woods: Ecological Changes and Peasant Resistance in the Himalaya* (Berkeley: University of California Press, 2000), 196.

THIRTEEN Conclusion

1. Alan Watts, *Nature, Man, and Woman* (New York: Vintage Books, 1991), 68–69.

2. Johann Wolfgang von Goethe, *Faust, Part Two* (1832; repr. Oxford: Oxford University Press, 1994), 234.

Index

Note: Figures and tables are indicated by "f" or "t," respectively, following page numbers.

Abercrombie, Nicholas, 93
Abrahamson, Mark, 220
Adams, S. D., 232–33
adaptation, 14
adaptive management, 211, 213–14, 235–37
Adhikari, Jagannath, 258
Adirondack Park, 33, 34
adopter curve, 31
age, 93, 162
age cycles, 36–37
air, 78
airsheds, 78
albedo, 78
Alihan, Milla Aïssa, 65–66
Allen, Timothy, 49
Amazon basin, 242–45
American dream, 248
American Sociology School, 69
anthropogenic ecosystems, 56
anthropology, 48, 69–70
armed institutions, 90

Armstrong, Karen, 84
Army Corps of Engineers, 42
art and crafts, 83, 159
Assateague Island National Seashore, 174
Audubon, John J., 25

Bailey, Kenneth D., 67
Baldwin, J. M., 32
Baltimore, Maryland, 148–209; assets of, 149; background of HEM project in, 72–73, 151–54; critical resources for, 154–60; HEM variables for, 155f; impoverished neighborhoods in, 226; interventions in, 163–65; lessons learned from, 168–209; outcomes in, 164–66; patch analysis of, 220; patterns of development in, 234–35; problems of, 148–49; Vital Signs (neighborhood indicators) of, 196–206t, 237; watershed, 169
Baltimore Ecosystem Study, 151, 162, 178, 207, 232

Baltimore Harbor Waterkeeper, 164
Baltimore Harbor Watershed Association, 164
Baltimore Neighborhood Indicators Alliance, 177–78, 209
Bangkok, 73
Bangladesh, 226
Barthes, Roland, 272n16
Bartram, William, 25
Bates, Marston, 69
beliefs, 83
Bell, Alma, 152
Bell, W., 217
Bennett, J. W., 69–70, 70f, 100
Berger, James O., 69
Berger, John, 246–47
Bhutan, 251–66
biogeochemical cycles, 18t
biology: ecosystem concept in, 51–63; and human ecosystems, 47; social sciences in relation to, 52–54, 64–66, 71
biomass accumulation model, 27
biophysical resources: in Baltimore case study, 154–57, 175, 179–80t; connected with Deepwater Horizon oil spill, 136t; as element of HEM, 19; use of, by the wealthy, 241–46; variables of, 76–79
biopiracy, 242–43
biotic communities, 54–55
black swans, 104
Bloods, 95
Blue Water Baltimore, 164, 165
boomer generation, 36, 212–13
Bormann, Bernard, 213, 235–36
Bormann, Frederick H., 27–28, 40–41, 52, 60, 154
Borneo, 3
Boston, Massachusetts, 220, 230
Bramwell, Anna, 231
Buddhism, 251–52, 260–63, 265
Burch, William, 24–25, 28, 42, 48, 71–73, 113, 151, 285n2

Burgess, Ernest W., 64, 217
Bush, George W., 164

Cadenasso, Mary L., 212–13, 218–19, 223, 237, 238
Callenbach, Ernest, 14, 18
Calvinism, 251
Canada, 2
capital, 81, 96, 163
capitalism, 250–51
Carey Institute of Ecological Studies, 207
Carey Murray Outdoor Education Center, 161
Caribbean, 3
Carrera, Jacqueline M., 164
carrying capacity, 53
cascading consequences: in Deepwater Horizon oil spill, 138–39; HEM model for, 99f; in human ecosystems, 100–101; in Hurricane Sandy, 144f, 142–45; in Mayan civilization collapse, 98, 100
caste, 92, 93
Catholicism, 87
Catton, W. R., 53
causality, 29, 155, 156f, 157, 186
Chesapeake Bay, 153, 154, 158, 159, 163
Chicago, Illinois: demographic flux in, 218–19; heat wave deaths in, 111–12, 221–22; informal norms in, 232
Chicago School, 64–66, 216–17
China, 2–3, 77
Chipko, 266
Christianity, 263
"The City and its Environment" seminar, 48
clans, 93
Claritias, 228
class, 93
Clements, Frederic, 52, 54–56, 66
Clinton, Bill, 113–14

clustering, 110, 150, 186, 215–18, 226–27, 229, 231, 233
Cohen, Patricia, 242
Cohen, Yehudi, 69
coherence, 12–13
commerce, 87–88, 197t, 200t, 203t
complexity: challenges presented by, 30; of ecosystems, 52; of human ecosystems, 106
complexity theory, 233
complex organism, 55
conceptual frameworks, 24–25
conceptual models, 24–25
Conoco Corporation, 48
conservation: in Bhutan and Nepal, 252–66; meanings of, 33–34
consumption, 1–2, 247–49
contingency, 14
Conway, Gordon, 39
Conze, Edward, 262–63
Cook, Robert E., 57
Cooke, Sam, 22
cooking, 5–6
Cooley, C. H., 32
coral reefs, 3
corrective justice, 86
Cottrell, W. F., 54, 76, 108–10, 112
counties, as unit of analysis, 117
coupled natural-human systems, 71, 134, 136–37, 143–46
Cowles, Henry C., 52, 55
crafts. *See* art and crafts
Crips, 95
critical resources: in Baltimore case study, 154–60, 193t; case examples of communities dependent on, 34–39; components and variables of, 19, 76–84; as element of human ecosystem, 13; indigenous vs. imported, 19; varying definitions of, 35
Cronon, William, 211, 217–18
cross-species comparisons, 228–29, 231

cultural ecology, 69–70, 70f, 87
cultural resources: in Baltimore case study, 158–60, 175; as element of HEM, 19; future impact of, 249–51; variables of, 13, 82–84
cycles: age, 36–37; analysis of, 36; in Baltimore case study, 161, 185, 196t, 200t, 203t; defined, 20; in social species, 229; types of, 90–92; variables of, 13

Dalton, Shawn, 158
Darrenbacher, Bill, 233
decision making: process of, 26, 106–7; significance of, 49
Deepwater Horizon (MS252) oil spill, 4, 10, 73, 133–41; biophysical resources connected with, 136t; cascading consequences in, 138–39; HEM model for, 135f; scenario building for, 138–41; scenario conceptual framework for, 137–38, 137f
deforestation, 3
Degler, Carl, 228
demography, 211, 214, 218–24
Department of Public Works, 164
Department of Recreation and Parks (Baltimore), 151–52, 157–64, 166, 169, 285n2; Strategic Action Plan, 175
Dewey, John, 32
Dhaka, 73
Diamond, Jared, 31, 100
Dice, L. R., 67–68
diesel technology, 108–10
diffusion theory, 30–32
distributive justice, 86
diversity, 218
division of labor, 223
Doby, John, 228
DOI. *See* U.S. Department of Interior
Doob, Leonard, 48
Duncan, O. D., 54, 65
Dunham, H., 65

Durkheim, Émile, 53, 223–28, 233, 237–38
dynamic equilibrium, 45f, 56, 96

Earls, Felton, 232
East, W. G., 71
ecology: Buddhism in relation to, 262–63; social sciences and, 51, 52, 216–17. *See also* human ecology
Ecology (journal), 57
economics, 30–31
ecosystem: complexity of, 52; defining, 54, 57, 62; development of concept of, 51–63; human role in, 62–63; machine analogy for, 61–63. *See also* environment; human ecosystems
ecosystem management: aesthetic aspect of, 49; overview of, 7–8; in Pacific Northwest, 114–17, 118–32t
education, 88, 160, 198t, 201–2t, 205t
Ehrlich, Anne, 2, 4, 98, 100
Ehrlich, Paul, 1, 2, 4, 27, 98, 100, 273n3
Ellen, Roy, 70
Ellison, Larry, 244–45
emergent structure, 60
Endangered Species Act, 87, 89
energy, 76, 96
entropy, 18
environment: challenges to, 1–4, 23–24, 241; cycles of, 92; interactions in, 3, 28–30, 211; responses to, 4; urban, indicators of, 198t, 201t, 204t. *See also* ecosystem
environmental anthropology. *See* anthropology
environmental cycles, 92, 161–62
environmental sociology, 54
epidemiology, 47
European Union, 2
evaluation, means and standards of, 105–7
ex post facto experiments, 23–24, 42, 226, 253, 257, 265

faith (religion), 87, 241, 249–66
familiarity, 230
Faris, R., 65
fauna, 79
Faust (character), vii–viii, 4
Faust (Goethe), vii–viii, 11, 269
Faust tales, vii–viii, 11, 27, 269
Federation of Community Forestry Users Nepal, 252, 259
feedback loops, 97
Fernández-Armesto, Felipe, 5–6
Fernow, Bernard, 33–34
Firey, Walter, 26–27, 49, 54, 217, 219
Five Rivers Landscape Management Project, 235–36
flora, 79, 198t, 201t, 204t
flows, 21, 96–97, 162–63
folk crimes, 94
food, 5–6
food-cycle relationships, 58f
Forbes, Stephen, 52
Force, Jo Ellen, 113–17
Ford, Henry, 243, 247
Fordlandia, 243–44
forest management, 37–39
Forest Summit, 113–14
formal norms, 94, 162
Fort McHenry National Monument and National Historic Shrine, 149, 170, 174–76
fracking, 24
Francis, Pope, 87
Freeman, Harold, 226
Freud, Sigmund, 52

Gans, Herbert, 230
gated communities, 225–26
Gautama, Siddhartha, 260
gender, 93, 162
geographic information system (GIS), 150, 158, 172
geography, 69, 71

Geronimus, Arline, 226
Gettysburg National Military Park, 174
goals, 103
Goethe, Johann Wolfgang von, vii–viii, 11, 102, 268–69
Goffman, Erving, 230–31
Golley, Frank, 29, 52, 54, 57, 154
Gordon, John, 152
Gould, Stephen Jay, 9, 51, 59
governance, 89
Graham, Steffi, 159
Great Britain, 3–4
greenhouse gases, 2
Grove, Morgan, 72, 166, 227, 237
Guha, Ram, 266
Gulf of Mexico, 133–41
Gwynns Falls Greenway, 159, 161, 163, 164, 173, 174
Gwynns Falls Park, Baltimore, 171, 173, 174, 238
Gwynns Falls watershed, Baltimore, 153, 154, 169
Gwynns Falls Watershed Association, 164, 173

Hagen, J., 54, 57
Hall, E. T., 231
Hansen, Richard, 100
Hays, Samuel P., 37
health, 86
heat wave, 111–12
HEM. *See* Human Ecosystem Model
Herring Run watershed, Baltimore, 153, 154, 169
Herring Run Watershed Association, 164
Heylyn, Peter, 8
hierarchies, social, 20, 94–95
Himalayan environment, 254–66
Hinduism, 251–52, 257–58, 265
Hoekstra, Thomas, 49
Homeland neighborhood, Baltimore, 176, 187, 190–91t, 192, 193–94t, 200–202t

horses, 46
Hoyt, Homer, 217
Hubbard Brook ecosystem studies, 27–30, 36, 40–42, 60, 153
Hughes, J. Donald, 87
human ecology: Bennett's paradigm of, 69–70, 70f; nature–nurture issue and, 53–54; roots of, 64–73, 216–17; Tansley and, 55–57
Human Ecosystem Model (HEM), 4; Baltimore as case study in, 148–209; cascading consequences of, 98, 99f, 100–101; case examples of, 31–32, 33–34, 43–44, 46; components and variables of, 74–101, 75f, 99f; conceptual framework of, 25; data utilization facilitated by, 170–71, 177–78, 195; as decision-making tool, 107; development of, 41, 48–49, 71–73, 114, 267–68; and energy resources study, 42–46; equilibrium version of, 115f; extending, 210–39; goals, strategies, tactics, and methods in, 103–5; historical precedents for, 23–24, 34–39, 51–63; methodological perspectives of, 47–48; overview of, 8–9, 12–22, 75f; population as driver variable of, 218–28; science-based assessments during environmental crises using, 133–47; SSG reliance on, 5; territory as driver variable of, 228–33; theories/models underlying, 26–33, 39–42, 69–70; usefulness and applications of, 5, 7, 27, 49–50, 72–73, 113–209, 268
human ecosystems: base conditions of, 15, 17–19; complexity of, 106; constants in, 15, 18; defined, 12–15; development of concept of, 67–68; elements of, 13; evolution of, 13, 100; flows in, 21; future challenges to, 240–69; Himalayan, 255–56; illustration of, 16f; interactions

human ecosystems (*continued*)
in, 22; machine analogy for, 13; O'Neill and, 63; open nature of, 13; organic nature of, 13; patch analysis and, 218; social system within, 45f; spatial scales of, 15; structure and dynamics of, 8–9; as theoretical framework, 113; time frames of, 14–15; variables in, 17. *See also* ecosystem
Hurricane Sandy, 4–5, 10, 73, 142–47; cascading consequences in, 142f, 142–45; scenario building for, 142–44
Hurricane Sandy Rebuilding Task Force, 142, 146
Hutchinson, G. Evelyn, 57, 214

identity, 92–93, 162
India, 2
individual cycles, 91
individuals, flows of, 96, 162
informal norms, 93–94, 162, 232
information, 80, 96, 158, 162–63, 246
inheritance of property, 86
institutional cycles, 91–92, 161–62
interdisciplinarity: obstacles to, 8, 48–49; required by environmental challenges, 3, 4, 10, 48; required by urban ecosystems, 208–9; SSG use of, 5
Intergovernmental Panel on Climate Change, 139
Interior Columbia Basin Ecosystem Management Project (ICBEMP), 114–17, 118–32t
International Council for Science, 72
IPAT model, 67
Islam, 263
Israel, 95

Jackson, Joe, 244
Jacobs, Jane, 224–25, 227, 230, 233
Johns Hopkins University, 149
Johnson, Steve, 224–25, 227
Jones, Ralph, 151–52, 160

Jones Falls watershed, Baltimore, 153, 154, 164, 169
Jones Falls Watershed Association, 164
justice, 86–87, 160, 241

Katahdin Woods and Waters National Monument, 34
Kaufman, Harold and Lois, 34, 37–39
Keller, A. G., 53
Kerner, Anton Joseph, 52
Key, Francis Scott, 149, 175
Kids Grow, 158, 160, 164, 166
Kiester, A. Ross, 235–36
kinships, 85–86
Klinenberg, Eric, 111–12, 219, 221–22, 230
knowledge, 95, 162, 246–47
Kornblum, William, 218–22, 230
Krishnamurti, Jiddu, 268
Kuchka, H. E., 71
Kuhn, T. S., 62

labor, 80–81, 158–59
land, 76–77
Landau, Misia, 230
Landis, Paul, 34–37
land use, 26–27, 40–41, 76–77
language, meaning of, 32–33
law. *See* formal norms; justice
lawns, 233
Leakin Park, Baltimore, 171
Leech, Susan, 7–8
leisure, 88–89, 161
Leontief, Wassily, 52
Levins, R., 39
Ley, David, 216–17
life science, 8, 10–11
Likens, Gene, 27–28, 40–41, 52, 60, 154
Lincoln County, Montana, 37–39
Lindeman, Raymond, 57–59
Lindh, A. G., 39
loess soil deposits, 77
Long Term Ecological Research (LTER), 151, 162, 170, 171, 173, 207–8

Loyola University, 149
Ludwig, Daniel, 244, 247
Luzadis, Valerie, 71

machine analogy, 13, 61–63
Machlis, Gary, 48, 73, 113–17
MacIver, R. M., 65
Madison-Eastend neighborhood, Baltimore, 176, 186–87, 188–89t, 192, 193–94t, 196–99t
Malmberg, Torsten, 231–32
Margalef, Ramon, 60
Marlowe, Christopher, vii–viii, 4
materials, 78–79, 96
May, Robert M., 4
Mayan civilization, 98, 100
McCord, Colin, 226
McDougall, Harold A., 220
Mead, George Herbert, 32
mechanical solidarity, 223–27, 238
medicine, 69
Merton, Robert, 106–7
Mesabi Range, Minnesota, 35–36
methods, 105
Michel, Sally, 152
Michelson, William H., 217
Miller, Bennett, 42
Milsum, J. H., 39
Mintz, Sydney, 5
Moore, Harry, 40
Moran, E. F., 70, 113
MS252 oil spill. *See* Deepwater Horizon (MS252) oil spill
Mumford, Lewis, 40, 64
myths, 84, 160

National Science Foundation (NSF), 151, 162, 171, 207
natural areas, 65, 216–17
natural disasters, 110
natural resource sociology, 54
negative feedback, 97
neighborhoods, cohesion of, 225–26

Nepal, 251–66
New Haven, Connecticut, 220
Newton, Isaac, 51
New Zealand, 77
norms, 93–94, 162
Northwest Forest Plan, 236
NSF. *See* National Science Foundation
nutrients, 79, 96

Obama, Barack, 34, 142
Odum, E. P. (Eugene), 52, 54, 59–61, 67, 113
Odum, Howard W., 40, 52
Odum, H. T., 52, 60–61, 113
Ogburn, William F., 54
Olmsted, John D., 25
Olmsted legacy in Baltimore, 149, 160, 173
Olson, Sherry, 226, 234–35
O'Neill, Robert V., 61–63
ooze, 57–58
Operational Group Sandy (OGS), 142
organicism, 13, 57, 59
organic solidarity, 223–24, 227, 238
organization, 82–83, 159, 199t, 202t, 206t
Outward Bound, 158, 160
Ouyang, Zhiyun, 154

Pacific Northwest, 113–17, 118–32t
Palestine, 95
Park, Robert E., 64
park and conservation departments, 169–70
Parks and People Foundation, 149, 151, 158, 161, 164–66, 169
patch analysis: biological use of, 211; and demography, 218–28; and diversity, 218; explanation of, 186, 212; HEM and, 223, 228; of human ecosystems, 216–18; of urban areas, 212–13; urban planning and, 215–16
patches, social. *See* social patches
Perritt, Marlyn J., 163
Perry, Laura, 152

Philippines, 226
Philips, John, 54–55
physiological cycles, 91
Pickett, Stewart, 72, 212–13, 219, 223, 237
Pinchot, Gifford, 33
PISTOL model, 67
Plains Indians, 46
POET model, 66, 66f
Poggi, Gianfranco, 226–27
politics. *See* governance
Ponder, Stephen, 245
population, 80. *See also* demography
population growth, 1, 53, 80, 85, 223
population theory, 211
Porter, Charles, 263
positive feedback, 97
Potential Rating Index for Zip code Markets (PRIZM), 227–28
power, 94, 242
Presses, 69–70
prestige. *See* status
process, as crucial to ecosystem concept, 57
progress, 101
property, inheritance of, 86
property rights and institutions, 95, 241–46

Quimby, Roxanne, 34

race, 92–93
Rae, Douglas, 220
Raney, Don, 24
Rappaport, Roy, 69
Rapport, D. J., 154
Raudenbush, Robert, 232
Rawlings-Blake, Stephanie, 164
religion. *See* faith
resilience, 214–15
resource management, 67–68, 100
resources. *See* critical resources
Revitalizing Baltimore, 163–65, 173
Rose, Arnold, 32

Ross, F. A., 65
rural sociology, 30–31
Russia, 95

Salazar, Ken, 141, 142
Sampson, Robert, 221, 232
San Francisco, California, 245
sanitation, 197t, 201t, 204t
scenario-building technique, 138–41, 142–44
Schmoke, Kurt, 152
scientific uncertainty levels, 139, 140t
second-order causes, 29, 155, 156f, 157
security, 90
Selznick, Philip, 54
Shaw, Clifford, 65
Shenandoah National Park, 174
Shevky, E., 217
Simmel, Georg, 32
Simon, Julian, 27, 273n3
Singh, A., 154
Siuslaw National Forest, 235–36
SmartDraw, 139
Smil, Vaclav, 17–19
social anthropology. *See* anthropology
social Darwinism, 215
social ecology, 65
social indicators, 115–16, 118–32t
social institutions: in Baltimore case study, 183–85t; cycles of, 91–92; defined, 19–20; in energy systems, 43–44; in impoverished areas, 226; trends in, 241; variables of, 84–90
social norms, 93–94
social order: components and variables of, 92–96; defined, 20–21, 92
social patches: analysis of, 215–16; demographic study of, 218–21; dynamics of, 238; function of, 238; neighborhood-level, 225–26; organizational forms of, 223–24; resilience of, 214–15; territory of, 228–33
social roles, 43–44

social sciences: biology in relation to, 52–54, 64–66, 71; ecological concepts in, 51, 52, 216–17; and human ecology concept, 64–71; role of, in ecosystem management, 7, 34, 41, 42–43
social species, 229
social system: in Baltimore case study, 160–63, 193–94t; components and variables of, 19–21, 84–96; defined, 19; dynamic equilibrium model of, 45f; as element of human ecosystem, 13, 45f
sociobiology, 229
socioeconomic resources: in Baltimore case study, 158–59, 181–82t; as element of HEM, 19; variables of, 80–82
Sonoran Desert bioregion, 72
Sorokin, Pitirim, 53
Soul Stirrers, 22
soy farms, 245
species decline, 3–4
Spencer, Herbert, 42, 52
Speth, Gus, 3
spotted owl crisis, 113–14
SSG. *See* Strategic Sciences Group
status, 94–95, 162, 242
St. Clair, Lucas, 34
Steffen, Jerome, 248–49
Steiner, Frederick, 69
Strategic Sciences Group (SSG), 5, 133–41, 142–46
strategies, 102–7
structuralism, 272n16
structure, as set of functioning parts, 60
succession, 54, 55, 56, 66
Sumner, W. G., 53
sustainability: biophysical, 14; of human ecosystems, 14; sociocultural, 14
sustenance, 89, 241, 246–49
symbolic interaction theory, 30, 32–33

tactics, 102–7
Tansley, Arthur, 52, 54–57, 59, 63, 67, 71, 73

Taoism, 268
technology: social effects of, 108–10; as socioeconomic resource, 81–82
territorial studies, 211
territory, 95–96, 162, 228–33
textbooks, 59
third-order causes, 29, 155, 156f, 157
Thomas, William Isaac, 32, 65
Thrasher, Frederick Milton, 65
Thurow, Lester, 2
time, 14–15
timing cycles. *See* cycles
Timms, D., 215–16
trees and vegetation, 152–53, 232–33, 256–57
trust, 90, 221, 225

uncertainty, 139, 140t, 235–36
United Nations University, 73
United States, 2, 3, 248
United States, territory of, 95
universality, 5–6
University of Chicago. *See* Chicago School
Upper Columbia River Basin, 114–17
Urban Arts Institute, Baltimore, 159
urban ecosystems: adaptive strategies of, 227; learning from, 168–69, 195, 208–9; resilience of, 214–15
urban planning: patch analysis in, 212–13, 215–16; tools for, 150–51
Urban Resources Initiative (URI), 148, 150–54, 157–58, 160, 162–66, 170–72, 175, 285n2
U.S. Bureau of Land Management, 42, 114
U.S. Department of Interior (DOI), 5, 133, 146; Strategic Sciences Group, 133–41, 142–46
U.S. Environmental Protection Agency (EPA), 24, 42, 207
U.S. Federal Energy Department, 42
U.S. Fish and Wildlife Service, 42

U.S. Forest Service, 37–39, 42, 113–14, 166, 171, 207
U.S. National Park Service, 42, 149, 151, 174
utopian communities, 243–44

values, 83
Vital Signs (neighborhood indicators), 177–78, 196–206t, 237

Walker, Charles, 48
Wang, Rusong, 154
Ward, Lester, 52
warfare ecology, 90
Warner, W. Lloyd, 233
water, 77, 154–55, 156f
watersheds, 78, 154–55, 156f
Watts, Alan, 268
wealth, 94, 241–45

Weaver, W. W., 65
Webb, Walter Prescott, 248
Weber, Max, 32, 250–51
Weiss, Charles, 139, 227
Whitney, Gordon, 232–33
Wickham, Henry, 242–43
Wilson, E. O., 87, 96, 229–30
Wirth, Louis, 54, 65
World Summit on Sustainable Development, 72
Wright, R. G., 113

Yale School of Forestry and Environmental Studies, 152, 158, 285n2
Yale University, 207

Znaniecki, Florian, 65
Zorbaugh, Harvey, 65